For an instant, i world went out o

Sunita could almost see a line being drawn in the sands of time—this was the moment that separated before and after. She nearly took the coward's route, wondering if Frederick would swallow the lie that Amil was Sam's. Then she realized she could not, would not do that. "Yes. He is yours. Amil is your son."

Now she understood the origin of the phrase "a deafening silence." This one trolled the room, echoed in her ears until she wanted to shout. Instead she waited, saw his body freeze, saw the gamut of emotions cross his face, watched as it settled into an anger so ice cold a shiver rippled across her skin.

Panic twisted her insides—the die had been cast, and she knew now that whatever happened, life would never be the same.

Dear Reader,

This book was very personal to me as it is partly set in Mumbai, where my own grandparents lived and where I spent many of my childhood summers—I loved the Hanging Gardens and still remember climbing the Old Woman's Shoe. However, I think I should note that it's there my resemblance to Sunita ends. I am definitely not model material!

Sunita had a very brief mention in my previous book, *Claimed by the Wealthy Magnate*, and Prince Frederick had a big part in that book as well as a cameo appearance in *Rafael's Contract Bride*. So their story has been bubbling in my brain for a while. It's a story about the importance of being yourself, and being able to trust in yourself and in the power of love. None of those things were easy for Sunita and Frederick, and that meant they had an obstacle-ridden course to their happy-ever-after.

Happy reading!

Nina xx

Nina Milne has always dreamed of writing for Harlequin Romance—ever since she played libraries with her mother's stacks of Harlequin romances as a child. On her way to this dream Nina acquired an English degree, a hero of her own, three gorgeous children and—somehow!—an accountancy qualification. She lives in Brighton and has filled her house with stacks of books—her very own *real* library.

Books by Nina Milne

Harlequin Romance

The Derwent Family

Rafael's Contract Bride
The Earl's Snow-Kissed Proposal
Claimed by the Wealthy Magnate

Christmas Kisses with Her Boss

Visit the Author Profile page at Harlequin.com for more titles.

To the memory of my very lovely
"Nanni"—I still miss you.

CHAPTER ONE

August 15th—Online Celebrity News with April Fotherington

Who will be the new Lycander Princess? All bets are off!

It's official! Lady Kaitlin Derwent is no longer a contender for the position of Lycander Bride—the people's favourite aristo announced that her new squeeze for the foreseeable future is Daniel Harrington, CEO of Harrington's Legal Services.

Who'd have thought it?

Exit Lady Kaitlin!

So Prince Frederick, ruler of Lycander, is on the lookout for a new bride.

Who will it be?

Will it be the type of woman who graced his arm and his bed back in his playboy days, before the tragic death of his older brother and the scandalous death of Prince

Alphonse, his flamboyant father, in a house of ill repute propelled him to the throne?

FREDERICK II OF the House of Petrelli, Prince and Ruler of Lycander, stopped reading and pushed his screen across the ornate carved desk, resplendent with gilt—a royal gift from an English monarch of yore.

The phrase pounded his brain—*tragic death of his older brother*—but he forced his features to remain calm, and made himself focus on the man standing in front of him: Marcus Alrikson, his chief advisor. After all, he needed all the advice he could get.

'I don't understand what the problem is—this article is nothing more than a gossip fest. And it's old news.'

Marcus shook his head. '*That* is the problem. The article serves to remind the people of your past.'

'Don't you mean my sordid, scandalous and immoral past?' Might as well tell it like it is, he thought.

'If you like,' Marcus returned evenly. 'The bigger problem is that we both know you are holding on to the crown by your fingertips. The people did not want you on the throne because of your past—so any reminder causes damage.'

'I understand that.'

The all too familiar guilt twisted his insides—the people had wanted his brother on the throne. Axel had been born to this. He would have been the ideal ruler to bring prosperity and calm to the land after their father's turbulent rule.

But Axel was dead and buried—victim of a car crash that should have been Frederick's destiny. Frederick should have been in that car on his way to a State dinner; instead he'd asked Axel to step in and take his place and his big brother had—no questions asked. So Frederick had attended a party on board a glitzy yacht to celebrate a business deal…and Axel had died.

The dark secret tarnished Frederick's soul, weighted his conscience.

And now Lycander was stuck with the black sheep of the royal line and the people were threatening to revolt. Bleak determination hardened inside him. He would keep the crown safe, whatever the cost—he owed that at least to Axel's memory.

'So what do you suggest?'

'I suggest you find a new bride—someone like Lady Kaitlin. Your proposed alliance with Kaitlin was a popular one. It showed the people that you had decided to settle down with a suitable bride, that you'd changed—proof there would be no repeat of your father's disastrous marriages.'

'I *have* decided to settle down.' To bind himself to a lifestyle he'd once sworn to avoid and the formulation of a cold-blooded alliance undertaken for the sake of the throne. 'But Kaitlin is no longer an option—she has fallen in love with another man.'

Irritation sparked inside him. He wished Kaitlin well, but it was hard to believe that the cool, poised Lady Kaitlin had succumbed to so foolish an emotion.

'Which is not good news for Lycander.'

Marcus resumed pacing, each stride swallowing up a metre of the marble floor, taking him past yet another portrait of one of Frederick's ancestors.

'Kaitlin was the perfect bride—her background is impeccable and she reminded the people of Lycander brides of the past.'

Unlike the succession of actresses, models and gold diggers Frederick's father had married.

'The people loved her.'

Unlike you.

The unspoken words hovered in the air between them.

'I understand all this. But Kaitlin is history.'

'Yes. And right now the press is *focused* on your history. That article zones in on your former flames—the actresses, the socialites, the models. Giselle, Mariana, Sunita… Hell, this reporter, April, even tried to track them down.'

Frederick froze.

Sunita.

Images flashed across his mind; memory reached across the chasm of tragedy.

Sunita.

Shared laughter, sheer beauty, almond-shaped eyes of a brown that veered from tawny to light, dependent on her mood. The raven sheen of her silken hair, the glow of her skin, the lissom length of her legs.

Sunita.

The woman who had left him—the woman he'd allowed to go…

Without preamble, he pulled his netbook back towards him, eyes scanning the article.

But where is Sunita now?

This is where it becomes a little myste-rious.

Mere weeks after the end of her relation-ship with the Prince of Lycander—which, according to several sources, she ended abruptly—Sunita decided to 'take a break' from her highly lucrative modelling career to 'rediscover her roots'.

This involved a move to Mumbai, where her mother reportedly hailed from. But the trail ends there, and to all intents and purposes Sunita seems to have vanished.

'Frederick?' Marcus's voice pulled him from the article and he looked up to see his chief advisor's forehead crease into a frown. 'What is it?'

'Nothing.' Under the sceptical gaze Frederick shrugged. 'It just sounds unlike Sunita to give up her career.'

Sunita had been one of the most ambitious people he knew—had been defined by that ambition, had had her career aspirations and goals mapped out with well-lit beacons. The idea of her jacking it all in seemed far-fetched at best.

Marcus drummed his fingers on his thigh. 'Could her disappearance have anything to do with you?'

'No.'

'What happened?'

'We spent a few weeks together—she moved on.'

'*She* moved on?'

Damn. 'We moved on.'

'Why?'

Keep it together. This is history. 'She decided to call it a day as she'd garnered sufficient publicity from our connection.'

Marcus raised his eyebrows. 'So she used you for publicity?'

'Yes. To be fair, she was upfront about that from the start.'

More fool him for thinking she'd changed her mind as time had gone on. He'd believed their time together, the long conversations, the laughter, had meant something. Well, he'd been wrong. Sunita had been after publicity and then she'd cut and run. Yet there had been something in her expression that morning…a transitory shadow in her tawny eyes, an errant twist of her hands that had belied the glib words dropping from her lips. But he hadn't called her on it.

Enough! The past was over and did not bear dwelling on because—as he knew with soul-wrenching certainty—it could not be changed.

Marcus's dark blue eyes met his as he resumed pacing. 'So weeks after this publicity stunt she disappeared off the modelling scene? That doesn't make sense.'

It didn't. But it had nothing to do with him. Two years ago Sunita had affected him in ways he didn't want to remember. He'd missed her once she'd gone—an unheard-of weakness he'd knocked on the head and buried. Easy come, easy go. That was the Playboy Prince's motto. Sunita had gone—he'd accepted it. And then, mere months after her departure, Axel had died and his life had changed for ever.

'I'll look into it,' Marcus said. 'But right now

you need to focus on this list. Potential brides. A princess, a lady and a *marquesa*. Take your pick.'

Frederick accepted the piece of paper but didn't so much as glance down. 'What do you mean, "look into it"?'

'If there is any chance of potential scandal we need to shut it down now. So I plan to find Sunita before April Fotherington or any other reporter does.'

'Then what?'

'Then I'll send someone to talk to her. Or go myself.'

'No!' The refusal came with a vehemence that surprised him. However it had ended, his time with Sunita had marked something—his last moments of joy before catastrophe occurred, perhaps. He didn't want her life tainted...didn't want Marcus or his minions to find her if she didn't want to be found.

'It needs to be done.' Marcus leant forward, his hands on the edge of the desk. 'I understand you don't like it, but you can't take even the smallest risk that there is a scandal floating around out there. The crown is at stake. The throne is rocking, Frederick, and if it topples it will be a Humpty Dumpty scenario.'

Great! A Humpty Dumpty scenario—exactly what he needed. Of course he could choose to ignore the warning, but that would be foolish.

Marcus knew his stuff. The sensible option would be to allow Marcus to go ahead, investigate and deal with any problem. But for some reason every fibre of his being cavilled—dammit, stupid though it sounded, it wasn't the *honourable* thing to do.

A small mocking smile tilted his lips as he faced his chief advisor. Frederick of Lycander—man of honour. Axel would be proud of him. 'Fine. *I'll* check out Sunita.'

Marcus's blue eyes narrowed. 'With all due respect, that's nuts and you know it. The press will jump on it.'

'Then let them jump. I'm the boss and this is what's going to happen.'

'Why?'

'Because it's the right thing to do.' And for once he'd like to stand on a tiny wedge of the moral high ground. 'What would Axel have done? Sent you in to spy on a woman he'd dated?'

'Axel would never have got himself into a position where it was necessary.'

'Touché. But I have and I will deal with it.' His brain whirred as he thought it through. 'I can schedule a trip to Mumbai—I'd like to follow up on how the Schools for All project is rolling out anyway.'

It was a project set up by Axel, but Frederick had taken it over and had every intention of making it into a success.

'I'll locate Sunita, confirm there is no scandal, and then I'll come back and find a wife from your shortlist. No argument.' A mirthless smile touched his lips. 'Don't worry. I'll be discreet.'

August 17th, Mumbai

Sunita stared down at the screen and reread the article for approximately the millionth time in three days as a mini-tornado of panic whirled and soared around her tummy.

She told herself that she was climbing the heights of irrationality. April Fotherington *hadn't* found her—she was safe here in this spacious, anonymous Mumbai apartment, surrounded by cool white walls and the hustle and bustle of a city she'd come to love. Soon enough the flicker of interest the article might ignite would die out. No one had discovered her secret thus far—there was no reason to believe they would now. She was safe. *They* were safe.

But she couldn't help the sudden lurch of fear as she gazed round the living room and the evidence of the life she'd created. Signs of her baby son were everywhere—a wooden toy box in the corner, the cheerful play mat by the sofa, board books, beakers... She knew all too well how quickly life could change, be upended and destroyed.

Stop. No one would take Amil away. Alphonse of Lycander was dead, and he had been the greatest threat—a man who had fought virulent custody battles for four of his children and used his position and wealth to win them all. She had no doubt he would have done the same for his grandchild—would have used the might and power of his sovereignty to win Amil.

Just as Frederick still might.

The peal of the doorbell jolted her from her thoughts and a scud of panic skittered through her. It couldn't be her grandmother and Amil—they had only left a little while before. *Chill*. They could have forgotten something, it could be a neighbour, or a delivery or—

Only one way to find out.

Holding her breath, she peered through the peephole.

Shock dizzied her—she blinked and prayed the man at her door was a figment of her overheated imagination, brought on by reading the article so many times. The alternative was too ghastly to contemplate. But, however many times she blinked, Prince Frederick of Lycander was still right there.

What to do? What to do? Ignore him?

But what if he waited outside? What if he was

still there when Amil came back? Or what if he went away and returned when Amil was here? What if he was here to take Amil?

Enough. She had not got this far to give up now. She was no longer that ten-year-old girl, reeling from her mother's death, powerless to stop the father she had never known from taking her. No longer that eleven, twelve, thirteen-year-old girl at the mercy of her stepmother and sisters who had graduated with honours from *Cinderella* school.

She'd escaped them without the help of a handsome prince and left that feeling of powerlessness far behind. No way was she going back there—especially now, when her son was at stake.

Adrenalin surged through her body as she did what life had taught her—moved forward to face up to whatever was about to be thrown at her. She might dodge it, catch it, or punch it, but she would confront it on her own terms.

True to her motto, she pulled the door open and raised her eyebrows in aloof surprise. 'Your Highness,' she said. 'What are you doing here?'

Stepping out into the communal hall, she closed the door behind her, searching his gaze for a sign that he knew about Amil.

'I came to see *you*. April Fotherington wrote an article saying you'd vanished.'

Sunita forced herself not to lean back against the wall in relief. Instead, she maintained her façade of reserve as they stood and studied each other. Against her will, her stomach nosedived and her hormones cartwheeled. Memories of the totally wrong sort streamed through her mind and fizzed through her veins as she drank him in. The same corn-blond hair, the same hazel eyes...

No, not the same. His eyes were now haunted by shadows and his lips no longer turned upward in insouciance. Prince Frederick looked like a man who hadn't smiled in a while. Little wonder after the loss of his brother and his father, followed by a troubled ascent to the throne.

Instinctively she stepped closer, wanting to offer comfort. 'I saw the article. But before we discuss that, I'm sorry for your losses. I wanted to send condolences but...'

It had been too risky, and it had seemed wrong somehow—to send condolences whilst pregnant with his baby, whom she intended to keep secret from him.

'Why didn't you?'

The seemingly casual question held an edge and she tensed.

'If all your girlfriends had done that you'd still be reading them now. I didn't feel our brief relationship gave me the right.'

Disingenuous, but there was some truth there. For a second she could almost taste the bitter disappointment with herself for succumbing to the Playboy Prince's charms and falling into bed with him. Hell—she might as well have carved the notch on his four-poster bed herself.

She'd woken the morning after and known what she had to do—the only way forward to salvage some pride and dignity. End it on *her* terms, before he did. It had been the only option, but even as she had done it there had been a tiny part of her that had hoped he'd stop her, ask her to stay. But of course he hadn't. The Playboy Prince wouldn't change. People didn't change— Sunita knew that.

Anyway this was history. *Over and done with.*

'I am offering condolences now.'

'Thank you. But, as I said, that's not why I am here.'

'The article?'

'Yes. I'd like to talk—perhaps we could go inside.'

'No!' *Tone it down, Sunita.* 'This is my home, Frederick, my private sanctuary. I want to keep it that way.'

He eyed her for a moment and she forced herself to hold his gaze.

'Then where would you suggest? Preferably somewhere discreet.'

'In case the press spot us and tips me as the next candidate for Lycander Bride?'

The words were out before she could stop them; obscure hurt touched her with the knowledge he didn't want to be seen with her.

'Something like that. You're my unofficial business.'

For a moment there was a hint of the Frederick she'd known in the warmth of his voice, and more memories threatened to surface. Of warmth and laughter, touch and taste.

'My official reason for this trip is charity business—I'm patron of an educational charity that is rolling out some new schools.'

The tang of warmth had disappeared; instead impatience vibrated from him as he shifted from foot to foot.

'Are you sure we can't talk inside? It shouldn't take long. All I want is the solution to April's mystery.'

Sunita checked the hollow laughter before it could fall from her lips. Was that *all* he wanted? Easy-peasy, lemon-squeezy.

'I'm sure we can't talk here.'

Think. But coherent thought was nigh on impossible. Raw panic combined with her body's reaction to his proximity had unsettled her, sheer awareness wrong-footed her. *Think.* Yet her mind drew a blank as to any possible location, any café where she and Amil weren't regulars.

Fear displaced all other emotion—Frederick must not find out about Amil. Not now, not like this. One day, yes, but at a time of her choice—when it was right and safe for Amil.

'I'll just grab a coat and we can go.'

'A coat?'

'It's monsoon season.'

Sunita turned, opened the door, and slipped inside, her mind racing to formulate a plan. She'd always been able to think on her feet, after all. If Frederick wanted a solution to the mystery of her disappearance from the modelling scene, then that was what she would provide.

Grabbing her phone, she pressed speed dial and waited.

'Sunita?'

'Hey, Sam. I need a favour. A *big* favour.'

CHAPTER TWO

FREDERICK WATCHED AS she opened the door and
sidled out. Coatless, he couldn't help but notice.
What was going on? Anyone would think she
had the Lycander Crown Jewels tucked away in
there. Hell, maybe she did. Or maybe something
was wrong.

Disquiet flickered and he closed it down.
He'd vowed emotion would not come into play
here. He and Sunita were history—the sole reason for his presence was to ensure no scandal
would touch Lycander and topple him, Humpty
Dumpty-style.

They exited the building and emerged onto
the heat-soaked pavement, thronged with an almost impossible mass of people, alive with the
shouts of the hawkers who peddled their wares
and the thrum of the seemingly endless cars that
streamed along the road. Horns blared, and the
smell of cumin, coriander and myriad spices
mingled with the delicate scents of the garlands

of flowers on offer and the harsher fumes of pollution.

Sunita walked slightly ahead, and he took the opportunity to study her. The past two years had done nothing to detract from her beauty— her hair shone with a lustre that should have the manufacturer of whatever brand of shampoo she used banging at her door, and her impossibly long legs and slender waist were unchanged.

Yet there was a difference. The Sunita he'd known had dressed to be noticed, but today her outfit was simple and anonymous—cut-off jeans, a loose dark blue T-shirt and blue sandals. It was an ensemble that made her blend in with the crowd. Even the way she walked seemed altered—somehow different from the way she had once sashayed down the catwalk.

Once.

And therein lay the crux of the matter. The more he thought about it, the more he recalled the vibrant, publicity-loving, career-orientated Sunita he'd known, the less possible it seemed that she had traded the life path she'd planned for an anonymous existence. His research of the past two days had confirmed that mere weeks after Sunita had ended their association she'd thrown it all away and melted into obscurity.

'How did you find me?'

'It wasn't easy.'

Or so Marcus had informed him. Sunita's agent had refused point-blank to respond to his discreet enquiries, but Marcus had ways and means, and had eventually procured the address through 'contacts'—whatever that meant.

'Was it my agent? Was it Harvey?'

'No. But whoever it was I promise you they did you a favour.'

'Some favour.'

'You mean you aren't happy to see me?' he deadpanned.

A shadow of a smile threatened to touch her lips and he fought the urge to focus on those lips in more detail.

'Pass.'

Raising an arm, she hailed a taxi and they waited until the yellow and black vehicle had screeched through the traffic to stop by the kerb.

Once inside she leant forward to speak to the driver. 'Sunshine Café, please,' she said, and then sat back. 'I'm taking you to meet the solution to your mystery. The reason I stayed in India.'

Her eyes slid away from him for a fraction of a second and then back again as she inhaled an audible breath.

'His name is Sam Matthews. He used to be a photographer, but he's moved here and set up a beach café.'

'A boyfriend?'

Such a simple answer—Sunita had given it all up for love. A small stab of jealousy pierced his ribcage, caught him unawares. *Get real, Frederick*. So what if she walked straight into someone else's arms, into the real thing? That had never been his destiny. *Know your limitations. Easy come, easy go*. Two stellar life mottos.

'Yes.'

'Must be some boyfriend to have persuaded you to throw away your career. You told me once that nothing was more important to you than success.'

'I meant it at the time.'

'So you gave up stardom and lucre for love.'

A small smile touched her lips. 'Yes, I did.'

'And you're happy? Sam makes you happy?'

Her hands twisted on her lap in a small convulsive movement. She looked down as if in surprise, then back up as she nodded. 'Yes.'

A spectrum of emotion showed in her brown eyes—regret, guilt, defiance,—he couldn't settle on what it was, and then it was gone.

'I'm happy.'

Job done. Sunita had a boyfriend and she'd moved on with her life. There was no dangerous scandal to uncover. A simple case of over-vigilance from his chief advisor. He could stop the taxi now and return to his hotel.

Yet…something felt off. He could swear Sunita was watching him, assessing his reactions. Just like two years ago when she'd called it a day. Or maybe it was his own ego seeing spectres—perhaps he didn't *want* to believe another woman had ricocheted from him to perfect love. Sunita to Sam, Kaitlin to Daniel—there was a definite pattern emerging.

He glanced out of the window at the busy beach, scattered with parasols and bodies, as the taxi slowed to a halt.

'We're here,' she announced.

What the hell? He might as well meet this paragon who had upended Sunita's plans, her career, her life, in a way he had not.

Damn it. There was that hint of chagrin again. Not classy, Frederick. Not royal behaviour.

Minutes later they approached a glass-fronted restaurant nestled at the corner of a less populated section of sand, under the shade of two fronded palms. Once inside, Frederick absorbed the warm yet uncluttered feel achieved by the wooden floor, high exposed beam ceiling and polished wooden tables and slatted chairs. A long sweeping bar added to the ambience, as did the hum of conversation.

Sunita moved forward. 'Hey, Sam.'

Frederick studied the man who stood before them. There was more than a hint of wariness in his eyes and stance. Chestnut wavy hair, average height, average build, light brown eyes that returned his gaze with an answering assessment.

Sunita completed the introduction. 'Sam, Frederick—Frederick, Sam. Right, now that's done...'

'Perhaps you and I could have a drink and a catch-up? For old times' sake.'

The suggestion brought on by an instinctive unease, augmented by the look of reluctance on her face. Something wasn't right. She hadn't wanted him to so much as peek into her apartment, and she could have simply *told* him about Sam. Instead she'd brought him here to see him, as if to provide tangible proof of his existence.

'Sure.' Sunita glanced at her watch. 'But I can't be too long.'

Sam indicated a staircase. 'There's a private room you can use upstairs, if you want to chat without attracting attention.'

'Great. Thank you,' Frederick said, and stepped back to allow the couple to walk together.

Their body language indicated that they were...*comfortable* with each other. They walked side by side, but there was no accidental brush of a hand, no quick glance of appreciation or anticipation, no chemistry or any sign of the awareness that had shimmered in the air since he himself had set eyes on Sunita.

They entered a small room with a wooden table and chairs by a large glass window that overlooked the beach. Sam moved over to the window, closed the shutters and turned to face them. 'If you tell me what you'd like to drink, I'll have it sent up.'

'You're welcome to join us,' Frederick said smoothly, and saw the look of caution in Sam's brown eyes intensify as he shook his head.

'I'd love to, but we're extremely busy and one of my staff members didn't turn up today, so I'm afraid I can't.'

'That's fine, Sam. Don't worry,' Sunita interpolated—and surely the words had tumbled out just a little too fast. Like they did when she was nervous. 'Could I have a guava and pineapple juice, please?'

'Sounds good—I'll have the same.'

'No problem.'

With that, Sam left the room.

'He clearly doesn't see me as a threat,' Frederick observed.

'There is no reason why he should.'

For an instant he allowed his gaze to linger on her lips and he saw heat touch her cheekbones. 'Of course not,' he agreed smoothly.

Her eyes narrowed, and one sandaled foot tapped the floor with an impatience he remembered all too well. 'Anyway, you came here to solve the mystery. Mystery solved. So your "unofficial" business is over.'

Were her words almost too airy or had he caught a case of severe paranoia from Marcus? 'It would appear so.' He watched her from beneath lowered lids.

'So, tell me more about your official business—the schools project.'

'My brother set up the charity—he believed every child deserves access to an education, however basic.'

It had been a philanthropic side Frederick hadn't even known Axel had had—one his brother had kept private. Because he had been a good man…a good man who had died—

Grief and guilt thrust forward but he pushed them back. The only reparation he could make was to continue Axel's work.

'So, I'm funding and working with a committee to set up schools here. Tomorrow I'm going to visit one of the new ones and meet the children.'

'That sounds incredible—there's so much poverty here, and yet also such a vibrant sense of happiness as well.'

'Why don't you get involved? That would be great publicity for the organisation—I could put you in touch.'

For a second her face lit up, and then she

shook her head. 'No. I'm not modelling at the moment and…'

'I'm not suggesting you model. I'm suggesting you get involved with some charity work.'

'I…I don't want any publicity at the moment—'

'Why not?'

'I… Sam and I prefer our life to be out of the spotlight.'

This still didn't make sense. Sunita had thrived in the spotlight, been pulled to it like a moth to a flame. But before he could point that out, the door opened and a waitress appeared with a tray.

'Thank you.' Sunita smiled as the girl placed the drinks on the table, alongside a plate of snacks that looked to range from across the globe. Tiny pizzas topped with morsels of smoked salmon nestled next to crisp, succulent *pakora*, which sat alongside miniature burgers in minuscule buns. 'These look delicious.'

Once the waitress had exited, Frederick sampled a *pakora*, savoured the bite of the spice and the crunch of the batter around the soft potato underneath. 'These are delicious! Sam runs an excellent kitchen.'

'Yes—he and…he has made a real success of this place.'

'You must be proud of him.'

'Yes. Of course.'

'Are you involved with the restaurant?'

'No.'

He sipped his drink, with its refreshing contrast of sharp and sweet. 'So what do you do now? Do you have a job?'

'I…'

Fluster showed in the heat that crept along her cheekbones, the abrupt swirling of her drink, the over-careful selection of a snack.

'I'm a lady of leisure.' Her eyes dared him to challenge her, but he couldn't help it—a snort of disbelief emerged. Sunita had been a human dynamo, always on the go, abuzz with energy, ideas and vibrancy.

'For real?'

'Yes.' Now her fingers tapped on the table in irritation. 'Why not? I'm lucky enough that I can afford not to work—I pay my own way.'

An undercurrent of steel lined her words— one he remembered all too well. 'Just like you did two years ago.'

It had become a standing joke—she'd refused point-blank to let him pay for anything, had insisted they split every bill down the middle. The one time he'd been foolish enough to buy her a gift, she'd handed it back.

I don't like to feel beholden. It's my issue, not yours. Keep it for your next woman. I pay my own way.

Apparently she still did.

He raised his hands in a gesture of peace. 'Where you get your money from is none of my business. I just can't imagine you doing nothing all day.'

'That's not how it is. I have a very fulfilling life.'

'Doing what?'

'None of your business. You came here to find out why I disappeared. I've told you—I fell in love, I've settled down, and I want to live my life quietly.'

Instinct told him there was something askew with the portrait she painted. Tension showed in the tautness of her body; but perhaps that tension had nothing to do with him.

'My chief advisor will be relieved—he is worried there is some mystery around your disappearance that could damage me.'

For a fraction of a second her knuckles whitened around her glass and then her eyebrows rose in a quizzical curve. 'Isn't that a tad far-fetched? To say nothing of egotistically paranoid?'

'Possibly,' he agreed, matching her eyebrow for eyebrow. 'But it also seems extremely far-fetched to me that you walked away from your career.'

'Well, I did.'

Frederick waited, but it appeared Sunita felt that sufficed.

'So you confirm that your retreat and subsequent dramatic change of lifestyle have nothing to do with me?'

Her glance flickered away and then she laughed. 'We spent one night together two years ago. Do you *really* think that your charms, manifest though they were, were sufficient to make me change my life?'

Put like that, he had to admit it sounded arrogantly self-involved. And yet... 'We spent more than one night together, Sunita.'

A wave of her hand dismissed his comment. 'A publicity stunt—nothing more.'

'OK. Let's play it your way. I can just about buy it that those weeks were all about publicity for you, but what about that night? Was *that* all for publicity?'

These were the questions he should have asked two years ago.

Her gaze swept away from him. 'No. It wasn't. I didn't intend that night to happen.'

'Is that why you left?'

It was as though the years had rolled back—he could almost imagine that they were in that five-star hotel in Paris, where they'd played truant from the glitzy party they'd been supposed to be at. Attraction had finally taken over and—

Whoa! Reel it in, Frederick!

'Yes, that's why I left. I broke my own rules. By sleeping with you I became just another notch on your bedpost—another woman on the Playboy Prince's conveyor belt. That was never meant to happen.'

'That's not how it was.'

'That's *exactly* how it was.' Tawny eyes challenged him.

'And if I'd asked you to stay?'

'You didn't.'

Her voice was flat and who could blame her? The point was that he hadn't. Because it had been easier to believe that she'd never cared, to stick by his *easy come, easy go* motto.

'But this is all beside the point—there was never a future for us. People don't change.' Her voice held utter conviction. 'You were The Playboy Prince…'

'And *you* were very clear that you had no desire for a relationship because you wanted to focus on your career. Then, just weeks later, you met Sam and realised he was the one and your career was no longer important?' It was impossible for Frederick to keep the scepticism out of his voice.

'Yes, I did.'

'So you changed.'

'Love changes everything.'

Damn it—he'd stake his fortune on the sincerity in her voice, and there was that irrational nip of jealousy again.

'So, yes,' she continued, 'I met Sam and I decided to take a break, and the break has extended to a couple of years. Simple. No mystery. That's what you came here to discover.'

Now her tone had lost the fervour of truth—he was nearly sure of it.

'You promise?' The words were foolish, but he couldn't hold them back.

She nodded. 'I promise…'

He studied her expression, saw the hint of trouble in her eyes and in the twist of her fingers under the table.

'No scandal will break over Lycander.'

'Then my work here is done.'

Yet an odd reluctance pulled at him as he rose from the chair and looked down at her, sure now that there was *more* than a hint of trouble in her eyes. *Not his business.* She'd made a promise and he believed her. He had a country to run, a destiny to fulfil…

'I wish you well, Sunita. I'm glad you've found happiness.'

'I wish you well too.'

In one lithe movement she stood and stretched out a hand, caught his sleeve, stood on tiptoe and brushed his cheek with her lips. Memory slammed into him—her scent, the silken touch of her hair against his skin—and it took all his powers of self-control not to tug her into his arms. Instead, he forced his body to remain still, to accept the kiss in the spirit it was being given—whatever that might be—though he was pretty damn sure from the heat that touched her cheeks that she wasn't sure either.

'I…goodbye.' Once again her hands twisted together as she watched him.

'Goodbye, Sunita.'

He headed for the door, stopped at her audible intake of breath, half turned as she said his name.

'Yes?'

'It…it doesn't matter. It was good to see you again.'

That only confirmed that she had intended to say something else, but before he could respond Sam entered and glanced at them both. 'All OK?'

'Everything is fine.' Sunita's voice was over-bright now. 'Frederick is just leaving.'

Minutes later he was in a taxi, headed back to the hotel. But as the journey progressed doubts hustled and bustled and crowded his brain. Something was wrong. He had no idea what, and it most likely had nothing to do with him. Quite possibly he had the wrong end of the stick. Undoubtedly wisdom dictated that he should not get involved. Sunita was more than capable of looking out for herself, and she had Sam to turn to. But what if Sam was the problem?

Hell.

Leaning forward, he gave the driver Sunita's address.

Damn it all to hell and back! Sunita strode the length of her lounge and resisted the urge to kick a bright red bean bag across the room. Venting wouldn't stem the onrush, the sheer *onslaught* of guilt, the veritable tsunami of distaste with herself.

Why, why, *why* had he turned up? Not telling Frederick for two years had been hard enough— lying directly to Frederick's face was another ballgame altogether. Especially as it was a face that mirrored Amil's—the angle of his cheekbone, the colour of his eyes, the subtle nuances that couldn't be ignored.

The guilt kept rolling on in and her stride increased. *Focus.* Concentrate on all the sensible, logical justifications for her actions.

The decision to keep Amil a secret had been one of the toughest she had ever faced, but it was a decision she still believed to be right. She'd done her research: the Lycanders had a track record of winning custody of their children and hanging the mothers out to dry.

Frederick's father, Prince Alphonse, had fathered five children by four wives; his first wife had died, but he'd fought and won vicious custody battles against all the other three.

Ah, pointed out her conscience, *but Alphonse is dead, and in any case Frederick is Amil's father.*

But Frederick was also his father's son, and who knew what he might do? The scandal of an illegitimate baby was the last thing Lycander's Prince needed at this juncture, and she had no idea how he would react.

She didn't like any of the possible scenarios—from a custody battle to show his people that he looked after his own, to an outright and public rejection of Amil. Well, damn it, the first would happen over her dead body and the second made her shudder—because she knew exactly how awful that rejection felt and she wouldn't put Amil through it.

But the Frederick she'd seen today—would he be so callous?

She didn't know. Her thoughts were muddled by the vortex of emotion his arrival had evoked. Because something had warmed inside her, triggering a whole rush of feelings. Memories had swooped and soared, smothering her skin in desire. Flashes of his touch, of their shared joy and passion…all of that had upended any hope of rational thought or perspective. Just like two years before.

When she'd first met Frederick she'd expected to thoroughly dislike him; his reputation as a cutthroat businessman-cum-playboy had seen to that. But when he'd asked her to dinner she'd agreed to it for the publicity. And at that dinner he'd surprised her. At the next he'd surprised her even more, and somehow, as time had gone on, they had forged a connection—one she had tried oh, so hard to tell herself was nothing more than temporary friendship.

Hah!

And then there had been that *stupid* tug of attraction, which had eventually prevailed and overridden every rule she'd set herself.

Well, not this time.

To her relief the doorbell rang. Amil's arrival would put an end to all this.

She dashed to the door and pulled it open, a smile of welcome on her face. A smile that froze into a rictus of shock.

'Frederick?'

She didn't know why she'd posed it as a question, since it clearly *was* Frederick. Her brain scrambled for purchase and eventually found it as she moved to swing the door shut, to hustle him out.

Too late.

He stepped forward, glanced around the room, and she could almost see the penny begin to drop—slowly at first, as cursory curiosity morphed into deeper question.

'You have a baby?'

His hazel eyes widened in puzzlement, and a small frown creased his brow as he took another step into her sanctum. His gaze rested on each and every item of Amil's.

'Yes.' The word was a whisper—all she could manage as her tummy hollowed and she grasped the doorjamb with lifeless fingers.

'How old?'

Each syllable was ice-cold, edged with glass, and she nearly flinched. No, she would not be intimidated. Not here. Not now. What was done was done, and—rightly or wrongly—she knew that even if she could turn back time she would make the same decision.

'Fourteen months.'

'Girl or boy?'

'Boy.'

Each question, each answer, brought them closer and closer to the inevitable and her brain wouldn't function. Instead, all she could focus on was his face, on the dawn of emotion—wonder, anger, fear and surely hope too?

That last was so unexpected that it jolted her into further words. 'His name is Amil.'

'Amil,' he repeated.

He took another step forward and instinctively she moved as well, as if to protect the life she had built, putting herself between him and her home.

'Is he mine?'

For an instant it was if the world went out of focus. She could almost see a line being drawn in the sands of time—this was the instant that separated 'before' and 'after'. For one brief instant she nearly took the coward's route, wondered if he would swallow the lie that Amil was Sam's. Then she realised she could not, *would* not do that.

'Yes. He is yours. Amil is your son.'

Now she understood the origins of a deafening silence. This one rolled across the room, echoed in her ears until she wanted to shout. Instead she waited, saw his body freeze, saw the gamut of emotion cross his face, watched as it settled into an expression of anger so ice-cold a shiver rippled her skin.

Panic twisted her insides—the die had been cast and she knew that now, whatever happened, life would never be the same.

CHAPTER THREE

STAY STILL. FOCUS ON remaining still.

The room seemed to spin around him, the white walls a rotating blur, the floor tilting under his feet. Good thing he didn't suffer from seasickness. Emotions crashed into him, rebounded off the walls of his brain and the sides of his guts. His heart thudded his ribcage at the speed of insanity.

A child. A son. *His* child. *His* son.

Fourteen months old.

Fourteen months during which his son had had no father. Anger and pain twisted together. Frederick knew exactly what it was like to have no parent—his mother had abandoned him without compunction in return for a lump sum, a mansion and a yearly stipend that allowed her a life of luxury.

Easy come, easy go.

Yes, Frederick knew what it was like to know a parent was not there for him. The anger unfurled in him and solidified.

'My son,' he said slowly, and he couldn't keep the taut rage from his voice.

He saw Sunita's awareness of it, but she stepped forward right into the force field of his anger, tawny eyes fierce and fearless.

'*My* son,' she said.

Stop.

However angry he was, however furious he was, he had to think about the baby. About Amil. Memories of the horrendous custody battles his father had instigated crowded his mind—Stefan, Emerson, Barrett—his father had treated all his sons as possessions.

'*Our* son,' he said.

The knowledge was surreal, almost impossible to comprehend. But it was imperative that he kept in control—there was too much at stake here to let emotion override him. Time to shut emotion down, just as he had for two long years. Move it aside and deal with what had to be done.

'We need to talk.'

She hesitated and then nodded, moving forward to close the front door. She watched him warily, her hands twisted together, her tawny eyes wide.

'How do you know he's mine and not Sam's?'

The look she gave him was intended to wither. 'I'm not an idiot.'

'That is a questionable statement. But what

you *have* shown yourself to be is a liar. So you can hardly blame me for the question, or for wanting a better answer than that. How do you know?'

Her eyes narrowed in anger as she caught her lower lip in her teeth and then released it alongside a sigh. 'Sam isn't my boyfriend. He has a perfectly lovely girlfriend called Miranda and they live together. I asked him to fake it to try and explain to you why I left the modelling world.'

'Is there a boyfriend at all?'

She shook her head. 'No.'

So there had been no one since him. The thought provoked a caveman sort of satisfaction that had no place in this discussion. Sunita had deceived him to his face in order to hide his son from him—now was not the moment to give a damn about her relationship status. Apart from the fact that it meant Amil was his.

Hold it together, Frederick. Shelve the emotion...deal with the situation at hand.

'Why didn't you tell me?'

Sunita started to pace. Her stride reminded him of a caged animal.

'Because I was scared.'

Halting in front of him, she looked so beautiful it momentarily pierced his anger.

'I know how hard this must be for you, but please try to understand I was terrified.'

For an instant he believed her, but then he recalled her profession, her ability to play to the camera, and he swatted down the foolish fledgling impulse to show sympathy and emitted a snort of disbelief.

'Terrified of what? What did I ever do to make you fear me?'

The idea was abhorrent—he'd witnessed his father in action, his delight in the exertion of power, and he'd vowed never to engage in a similar manner. Thus he'd embarked on a life of pleasure instead.

'It wasn't that straightforward. When we split obviously I had no idea I was pregnant. I found out a few weeks later and I was in shock. I did intend to tell you, but I decided to wait until I got to twelve weeks. And then your brother died. I *couldn't* tell you then, so I decided to wait some more.'

Now her expression held no apology, and her eyes met his full-on.

'And?'

'And obviously there was a lot of press at the time about Lycander. I did some research, and it's all there—your father fought custody battles over every one of his children except Axel, and that was only because Axel's mother died before he could do so. Your mother never saw you again, his third wife fought for years before she won the right even to *see* her son, and wife number four lost her case because he managed to make out she was unfit and she had to publicly humiliate herself in order to be granted minimal visiting rights.'

'That was my father—not me.'

'Yes, but *you* had become the Lycander heir. Are you saying your father wouldn't have fought for custody of his grandson? Even if you'd wanted to, how could you have stopped him? More to the point, would you have cared enough to try?'

The words hit him like bullets. She hadn't believed he would fight for the well-being of their child. She'd thought he would stand back and watch Alphonse wrest his son away.

He shook his head. *Do you blame her?* asked a small voice. He'd been the Playboy Prince—he'd worked hard, played harder, and made it clear he had no wish for any emotional responsibilities.

'I would *never* have let my father take our child from you.' He knew first-hand what it felt like to grow up without a mother. All the Lycander children did.

'I couldn't take that risk. Plus, you didn't want to be a father—you'd made it more than clear that you had no wish for a relationship or a child.'

'Neither did you.'

His voice was even, non-accusatory, but she bristled anyway, tawny eyes flashing lasers.

'I changed.'

'But you didn't give me the chance to. Not at any point in the past two years. Even if you could justify your deceit to yourself when my father was alive, you could have told me after his death.'

His father's death had unleashed a fresh tumult of emotion to close down. He'd had to accept that he would now never forge a relationship with the man who had constantly put him down—the man who had never forgiven him for his mother's actions. And on a practical level it had pitchforked him into the nightmare scenario of ascension to the throne.

But none of that explained her continued deceit.

'I read the papers, Frederick. You have had enough to contend with in the past year to keep your throne—the revelation of a love-child with me would have finished you off. You were practically engaged to Lady Kaitlin.'

'So you want my *gratitude* for keeping my child a secret? You've persuaded yourself that you did it for me? Is that how you sleep at night?'

'I sleep fine at night. I did what I thought was right. I didn't want Amil to grow up knowing that he had been the reason his father lost his throne, or lost the woman he loved. That is too big a burden for any child.'

The words were rounded with utter certainty.

'That was not your decision to make. At any point. Regardless of the circumstances, you should have come to me as soon as you knew you were pregnant. Nothing should have stopped you. Not Axel, not my father, not Kaitlin—*nothing*. You have deprived him of his father.'

'I chose depriving him of his father over depriving him of his mother.' Her arms dropped to her sides and a sudden weariness slumped her shoulders. 'We can argue about this for ever—I did what I thought was best. For Amil.'

'*And* you.'

'If you like. But in this case the two were synonymous. He needs me.'

'I get that.'

He'd have settled for any mother—had lived in hope that one of the series of stepmothers would give a damn. Until he'd worked out there was little point getting attached, as his father quite simply got rid of each and every one.

'But Amil also needs his father. That would be me.'

'I accept that you are his father.'

Although she didn't look happy about it, her eyes were full of wariness.

'But whether he needs you or not depends on what you are offering him. If that isn't good for him then he doesn't need you. It makes no odds whether you are his father or not. The whole "blood is thicker than water" idea sucks.'

No argument there. 'I *will* be part of Amil's life.'

'It's not that easy.'

'It doesn't matter if it's easy.'

'Those are words. Words are meaningless. Exactly *how* would it work? You'll disguise yourself every so often and sneak over here to see him on "unofficial business" masked by your charity work? Or will you announce to your people that you have a love-child?'

Before he could answer there was a knock at the door and they both stilled.

'It's my grandmother...with Amil.' Panic touched her expression and she closed her eyes and inhaled deeply. 'I don't want my grandmother to know until we've worked out what to do.'

Frederick searched for words, tried to think, but the enormity of the moment had eclipsed his ability to rationalise. Instead fear came to the podium—he had a child, a son, and he was about to meet him.

What would he feel when he saw Amil?

The fear tasted ashen—what if he felt nothing?

What if he was like his mother and there was no instinctive love, merely an indifference that bordered on dislike? Or like his father, who had treated his sons as possessions, chess pieces in his petty power games?

If so, then he'd fake it—no matter what he did or didn't feel, he'd fake love until it became real.

He hauled in a deep breath and focused on Sunita's face. 'I'll leave as soon as you let them in. Ask your grandmother to look after Amil tonight. Then I'll come back and we can finish this discussion.'

Sunita nodded agreement and stepped forward.

His heart threatened to leave his ribcage and moisture sheened his neck as she pulled the door open.

A fleeting impression registered, of a tall, slender woman with silver hair pulled back in a bun, clad in a shimmering green and red sari, and then his gaze snagged on the little boy in her arms. Raven curls, chubby legs, a goofy smile for his mother.

Mine. My son.

Emotion slammed into him—so hard he almost recoiled and had to concentrate to stay steady. Fight or flight kicked in—half of him wanted to turn and run in sheer terror, the other half wanted to step forward and take his son, shield him from all and any harm.

'Nanni, this is an old friend of mine who's dropped in.'

'Good to meet you.' Somehow Frederick kept his voice even, forced himself to meet the older woman's alert gaze. He saw the small frown start to form on her brow and turned back to Sunita. 'It was great to see you again, Sunita. 'Til later.'

A last glance at his son—*his son*—and he walked away.

Sunita scooped Amil up and buried herself in his warmth and his scent. She held him so close that he wriggled in protest, so she lowered him to the ground and he crawled towards his play mat.

'Thank you for looking after him.'

'I enjoyed it immensely. And thank *you*, Sunita, for allowing me to be part of Amil's life. And yours.'

'Stop! I have told you—you don't need to thank me.'

Yet every time she did.

'Yes, I do. I was neither a good mother nor a good grandmother. You have given me a chance of redemption, and I appreciate that with all my heart.'

'We've been through this, Nanni; the past is the past and we're only looking forward.'

Her grandmother's marriage had been deeply unhappy—her husband had been an autocrat who had controlled every aspect of his family's life with an iron hand. When Sunita's mother had fallen pregnant by a British man who'd had no intent of standing by her, her father had insisted she be disowned.

Sunita could almost hear her mother's voice now: *'Suni, sweetheart, never, ever marry a man who can control you.'*

It was advice Sunita intended to take one step further—she had no plans to marry anyone, *ever*. Her father's marriage had been a misery of incompatibility, bitterness and blame—an imbroglio she'd been pitchforked into to live a Cinderella-like existence full of thoughtless, uncaring relations.

'Please, Nanni. You are a wonderful grandmother and great-grandmother and Amil adores you. Now, I have a favour to ask. Would you mind looking after Amil for the rest of the evening?'

'So you can see your friend again?'

'Yes.'

'The friend you didn't introduce?'

Sunita opened her mouth and closed it again.

Her grandmother shook her head. 'You don't have to tell me.'

'I *will* tell you, Nanni—but after dinner, if that's OK.'

'You will tell me whenever you are ready. Whatever it is, this time I will be there for you.'

An hour later, with Amil fed and his bag packed, Sunita gave her grandmother a hug. She watched as the driver she'd insisted on providing manoeuvred the car into the stream of traffic, waved, and then made her way back upstairs... To find the now familiar breadth of Frederick on the doorstep, a jacket hooked over his shoulder.

'Come in. Let's talk.'

He followed her inside and closed the door, draping his dark grey jacket over the back of a chair. 'Actually, I thought we could talk somewhere else. I've booked a table at Zeus.'

Located in one of Mumbai's most luxurious hotels, Zeus was the city's hottest restaurant, graced by celebrities and anyone who wanted to see and be seen.

Foreboding crept along her skin, every instinct on full alert. 'Why on earth would you do that?'

'Because I am taking the mother of my child out for dinner so we can discuss the future.'

Sunita stared at him as the surreal situation deepened into impossibility. 'If you and I go out for dinner it will galvanise a whole load of press interest.'

'That is the point. We are going public. I will not keep Amil a secret, or make him unofficial business.'

She blinked as her brain crashed and tried to change gear. 'But we haven't discussed this at all.'

This was going way too fast, and events were threatening to spiral out of control. *Her* control.

'I don't think we should go public until we've worked out the practical implications—until we have a plan.'

'Not possible. People are already wondering where I am. Especially my chief advisor. People may have spotted us at the café, and April Fotherington will be wondering if my presence in Mumbai is connected to you. I want the truth to come out on *my* terms, not hers, or those of whichever reporter makes it their business to "expose" the story. I want this to break in a positive way.'

Sunita eyed him, part of her impressed by the sheer strength and absolute assurance he projected, another part wary of the fact he seemed to have taken control of the situation without so much as a by-your-leave.

'I'm not sure that's possible. Think about the scandal—your people won't like this.' And they wouldn't like *her*, a supermodel with a dubious past. 'Are you sure this is the best way to introduce Amil's existence to your people?'

'I don't know. But I believe it's the right way to show my people that this is *good* news, that Amil is not a secret. That I am being honest.'

An unpleasant twinge of guilt pinched her nerves—*she* had kept Amil secret, *she* had been dishonest. She had made a decision that no longer felt anywhere near as right as it had this morning.

'So what do you say?' he asked. 'Will you have dinner with me?'

The idea gave her a sudden little thrill, brought back a sea of memories of the dinners they had shared two years before—dinners when banter and serious talk had flown back and forth, when each word, each gesture, had been a movement in the ancient dance of courtship. A courtship she had never meant to consummate…

But this meal would be on a whole new level and courtship would not be on the table. Wherever they held this discussion tonight, the only topic of conversation would be Amil and the future.

And if Frederick believed this strategy was the best way forward then she owed him her co-operation.

'Let's get this show on the road.' An unexpected fizz of excitement buzzed through her.

She could *do* this; she'd always relished a fight and once upon a time she'd revelled in a show. 'But I need to change.'

'You look fine to me.'

His voice was deep and molten, and just like that the atmosphere changed. Awareness hummed and vibrated, shimmering around them, and she had to force herself to remain still, to keep her feet rooted to the cool tiles of the floor. The hazel of his eyes had darkened in a way she remembered all too well, and now it was exhilaration of a different sort that heated her veins.

Stop.

All that mattered here was Amil and his future. Two years ago she had tried and failed to resist the magnetic pull that Frederick exerted on her—a pull she had distrusted, and this time would not permit. Whatever her treacherous hormones seemed to think.

Perhaps he realised the same, because he stepped backwards and nodded. 'But I appreciate you want to change.'

'I do. You need a show, and a show is exactly what I can provide. Luckily I kept some of the clothes from my modelling days.'

Even if she'd never once worn them, she loved them still. Silk, chiffon and lace, denim and velvet, long skirts and short, flared and skinny—

she had enjoyed showcasing each and every outfit. Had refused to wear any item that didn't make her soul sing. And now there was no denying the buzz. This was what she had once lived for and craved. Publicity, notice, fame— all things she could spin and control.

Almost against her will, her mind fizzed with possibility. Amil was no longer a secret, no longer in danger—they could live their lives as they wished. She could resume her career, be Sunita again, walk the catwalks and revel in fashion and all its glorious aspects. Amil would, of course, come with her—just as she had accompanied her mother to fashion shoots—and Nanni could come too.

Life would take on a new hue without the terrible burden of discovery clouding every horizon. Though of course Frederick would be part of that life, if only a minor part. His real life lay in Lycander, and she assumed he would want only a few visits a year perhaps.

Whoa! Slow right down, Suni!

She had no idea what Frederick's plans were, and she'd do well to remember that before she waltzed off into la-la land. She didn't know this man—this Frederick.

Her gaze rested on him, absorbed the breadth of his body, his masculine presence, the determined angle of his stubbled jaw, the shadowed eyes crinkled now in a network of lines she thought probably hadn't come from laughter. Her breath caught on a sudden wave of desire. Not only physical desire, but a stupid yearning to walk over and smooth the shadows away.

A yearning she filed away under both dangerous and delusional as she turned and left the room.

CHAPTER FOUR

FREDERICK CHANGED INTO the suit he'd had delivered to him whilst he was waiting and prowled the flat on the lookout for evidence of Amil's life.

Amil. The syllables were still so unfamiliar—his only knowledge of his son that brief glimpse a few hours earlier. But there would be time—plenty of time—to catch up on the past fourteen months. Provided Sunita agreed to his proposition—and she *would* agree.

Whatever it took, he would make her see his option was the only way forward.

He paused in front of a framed photograph of Sunita and a newborn Amil. He looked at the tiny baby, with his downy dark hair, the impossible perfection of his minuscule fingernails, and the utter vulnerability of him twisted Frederick's gut.

Shifting his gaze to Sunita, he saw the love in her brown eyes clear in every nuance, every part of her body. Her beauty was unquestionable, but this was a beauty that had nothing to do with physical features and everything to do with love.

Perhaps he should feel anger that he had missed out on that moment, but his overwhelming emotion was relief—gratitude, even—that his son had been given something so vital. Something he himself had never received. *His* mother had handed him straight over to a nanny and a few scant years later had disappeared from his life.

For a long moment Frederick gazed at the photo, trying to figure out what he should feel, what he *would* feel when he finally met Amil properly, held him… Panic hammered his chest and he stepped backwards. What if he was like his mother—what if he quite simply lacked the parenting gene?

The click of heels against marble snapped him to attention and he stepped back from the photo, turning to see Sunita advance into the room. For a moment his lungs truly ceased to work as his pulse ratcheted up a notch or three.

Sunita looked… It was impossible to describe her without recourse to a thesaurus. *This* was the woman he remembered—the one who dressed to catch the eye. But it wasn't only the dress with its bright red bodice and gently plumed skirt that showcased her trademark legs. The bright colour was toned down by contrasting black satin panels and silver stiletto heels. It was the way she wore it—she seemed to bring the dress alive. And vice versa. A buzz vibrated from her—an energy and sparkle that epitomised Sunita.

'Wow!' was the best he could do as he fought down visceral desire and the need to tug her into his arms and rekindle the spark that he knew with gut-wrenching certainty would burst into flame. To kiss her senseless...

What the hell was he thinking? More to the point, what part of himself was he thinking *with*?

Maybe he was more like his father than he knew. Alphonse had always put physical desire above all else. If he'd been attracted to a woman he'd acted on that attraction, regardless of marriage vows, fidelity or the tenets of plain, common decency. The last ruler of Lycander had believed that *his* desires were paramount, and it didn't matter who got hurt in the process.

Frederick wouldn't walk that road. He never had—that, at least, was one immoral path he'd avoided.

His business with Sunita was exactly that—*business*. He had an idea to propound, an idea he would not mix with the physical.

'You look fantastic.'

'Thank you. I know it sounds shallow, but it is awesome to dress up again.'

She smoothed her hand down the skirt and her smile caught at his chest.

'You look pretty good yourself. Where did the suit come from?'

'I had it delivered whilst I was waiting.'

'Good thinking, Batman.'

Her voice was a little breathless, and he knew that she was as affected as he was by their proximity. Her scent teased him, her eyes met his, and what he saw in their deep brown depths made him almost groan aloud.

Enough.

Right now he had to focus on the most important factor, and that was Amil. Irritation scoured him that he could be letting physical attraction come into play.

He nodded to the door. 'We'd better go.'

Sunita wanted, *needed* this journey to come to an end. Despite the spacious interior of the limo, Frederick was too…*close.*

Memories lingered in the air, and her body was on high alert, tuned in to his every move, and she loathed her own weakness as much now as she had two years before. She needed to distract herself, to focus on what was important—and that was Amil.

The day's events had moved at warp speed and she was desperately trying to keep up. The truth was out, and it was imperative she kept control of a future that she could no longer reliably predict.

Frederick wanted to be a real part of Amil's life—he had made that more than clear. But at this point she had no idea what that meant, and she knew she had to tread carefully.

The limo slowed down and she took a deep breath as it glided to a stop.

'Ready?' he asked.

'Ready.'

With any luck she wouldn't have lost her touch with the press. In truth, she'd always liked the paparazzi. Her mother had always told her that publicity was a means to measure success, part of the climb to fame and fortune and independence.

They stepped out into a crowd of reporters, the click of cameras and a fire of questions.

'Are you back together?'

'Friends or lovers?'

'Does Kaitlin know?'

'Where have you been, Sunita?'

Frederick showed no sign of tension. His posture and smile were relaxed, his whole attitude laid back.

'At present we have no comment. But if you hold on I promise we will have an announcement to make after dinner.'

Next to him, Sunita smiled the smile that had shot her to catwalk fame. She directed a small finger-wave at a reporter who'd always given her positive press, a smile at a woman she'd always enjoyed a good relationship with, and a wink at a photographer renowned for his audacity.

Then they left the reporters behind and entered the restaurant, and despite the knowledge of how important the forthcoming conversation was a part of Sunita revelled in the attention she was gathering. The simple ability to walk with her own natural grace, to know it was OK to be recognised, her appreciation of the dress and the inner confidence it gave her—all of it was such a contrast to the past two years, during which she had lived in constant denial of her own identity, burdened by the fear of discovery.

The manager beamed at them as he led them past the busy tables, where patrons looked up from their food and a buzz immediately spread. Sunita kept her eyes ahead, noting the dark-stained English Oak screens and latticing that graced the room, the hustle and bustle from the open-plan kitchen where chefs raced round, the waiters weaving in and out, and the tantalising smells that drifted into the eating area.

'As requested, we've seated you in a private dining area where you will be undisturbed. My head chef has arranged a buffet for you there, with samples of all our signature dishes, and there is, of course, champagne on ice—we are very happy to welcome you both here.'

He turned to Sunita.

'I do not expect you to remember me, but when you were a child your mother brought you many, many times to the restaurant I worked in then. Your mother was a lovely lady.'

Memory tugged as she studied the manager's face. 'I *do* remember you. You're Nikhil! You used to give me extra sweets and fortune cookies, and you would help me read the fortunes.'

His smile broadened. 'That is correct—I am very happy to see you here, and I am very sorry about your mother. She was a good woman.'

'Thank you. That means a lot to me. And it would have to her as well.'

It really would. So many people had looked down on Leela Baswani because she had been a single, unmarried mother, and a model and actress to boot. But her mother had refused to cower before them; she had lived her life and she'd loved every minute of it—even those terrible last few months. Months she didn't want to remember, of watching her mother decline, knowing that soon she would be left alone in the world.

But those were not the memories Leela would have wanted her daughter to carry forth into life.

Instead she would remember her as Nikhil did—as a good, brave, vibrant woman.

Nikhil showed them into the private dining room, where a beautifully decorated table laid with snowy white linen held fluted glassware, gleaming cutlery and a simple table decoration composed of an arrangement of glorious white roses.

Sunita looked at them, and then at Nikhil, and a lump formed in her throat. White roses had been her mother's favourite flower—her trademark accessory—and as the scent reached her nostrils she closed her eyes for a second. 'Thank you, Nikhil.'

The manager gave a small bow. 'You are very welcome. Now, both of you enjoy the food. I believe our chef has excelled himself. And I guarantee you complete privacy.'

With one more beaming smile, he left, closing the door behind him.

'I'm sorry about your mother,' Frederick said.

'Thank you.'

'She isn't mentioned in any articles about you except April's most recent one. None of your family is.'

'No. They aren't.'

And that was the way it would stay—she would love to remember her mother more publicly, but to do that would risk questions about her father, and she'd severed her ties with him years before—the man who'd abandoned her before birth and then reappeared in her life only to make it thoroughly miserable.

'Anyway, we aren't here to speak of my family.' He raised an eyebrow and she bit her lip. 'I mean, we are here to discuss Amil's future.'

'We are. But first shall we help ourselves to food?'

She nodded. No way could she hurt Nikhil's feelings, but she sensed there was more to Frederick's suggestion than that. It was almost as if he were stalling, giving himself time to prepare, and a sense of foreboding prickled her skin— one she did her best to shake off as she made a selection from the incredible dishes displayed on the table.

There was a tantalising array of dumplings with descriptions written in beautiful calligraphy next to each platter—prawn and chive, shanghai chicken, *pak choi*... Next to them lay main courses that made her mouth water—Szechuan clay pot chicken, salmon in Assam sauce, ginger fried rice...

The smell itself was enough to allay her fears, and she reminded herself that Frederick had a country to run—other fish to fry, so to speak. Surely the most he would want would be to contribute to Amil's upkeep and see him a few times a year. That would work—that would be more than enough.

Once they were seated, she took a deep breath. 'Before we start this discussion you need to know that I will not agree to anything that feels wrong for Amil. He is my priority here and if you try to take him away from me I will fight you with my last dying breath. I just want that out there.'

There was something almost speculative in his gaze, alongside a steely determination that matched her own. 'Amil is my priority too—and that means I *will* be a real part of his life. That is non-negotiable. *I* just want *that* out there.'

'Fine. But what does that mean?'

'I'm glad you asked that, because I've given this some thought and I know what I believe is best for Amil's future.'

The smoothness of his voice alerted Sunita's anxiety. The presentiment of doom returned and this time her very bones knew it was justified. Spearing a dumpling with an effort at nonchalance, she waved her fork in the air.

'Why don't you tell me what you have in mind?'

His hazel eyes met hers, his face neutral. 'I want you to marry me.'

CHAPTER FIVE

'Marry you?' Suita stared at him, flabberghasted. 'That's a joke, right?'

It must be his opening bid in negotiations designed to throw her into a state of incoherence. If so, he'd slammed the nail on the head.

'No joke. Trust me, marriage isn't a topic I'd kid about. It's a genuine proposal—I've thought it through.'

'When? In the past few hours? Are you certifiably *nuts*?'

'This makes sense.'

'How? There is no universe where this makes even a particle of a molecule of sense.'

'This is what is best for Amil—best for our son.'

'No, it isn't. Not in this day and age. You *cannot* play the let's-get-together-for-our-child's-sake card.'

That was the stuff of fairy tales, and she was damned sure that her mother had been right about those being a crock of manure.

'Yes, I can. In the circumstances.'

'What circumstances?' Her fogged brain attempted to illuminate a pathway to understanding and failed.

'If you marry me Amil will become Crown Prince of Lycander after me. If you don't, he won't.'

The words took the wind out of the sails of incredulity. Of *course*. *Duh!* But the idea that Frederick would marry her to legitimise Amil hadn't even tiptoed across her mind. The whole concept of her baby one day ruling a principality seemed surreal, and right now she needed to cling onto reality.

'We can't get married to give Amil a crown.'

'But we can get married so that we don't deprive him of one.'

'Semantics.' *Think.* 'He won't feel deprived of something he never expected to have.' *Would he?* 'Amil will grow up knowing…'

Her voice trailed off. Knowing what? That if his mother had agreed to marry his father he would have been a prince, a ruler, rather than a prince's illegitimate love-child.

'Knowing that he can be whatever he wants to be,' she concluded.

'As long as what he wants to be isn't Ruler of Lycander.'

Panic stole over her, wrapped her in tentacles

of anxiety. 'You are putting me in an impossible position. You are asking me to decide Amil's entire future. To make decisions on his behalf.'

'No. I am suggesting *we* make this decision together. I believe this is the best course of action for Amil. If you think otherwise then convince me.'

'He may not want to be pushed into a pre-ordained future—may not want to be a ruler. Why would we burden him with the weight of duty, with all the rules and obligations that come with it?'

'Because it is his *right* to rule. Just as it was my brother's.'

His voice was even, but she saw the shadows chase across his eyes, sensed the pain the words brought.

'Axel wanted to rule—he believed in his destiny.'

'So you believe this is Amil's destiny?' Sunita shook her head. 'It's too abstract. We make our own destiny and Amil will make his, whatever we decide to do. I want to make the decision that is best for his wellbeing and happiness—you don't need a crown for either.'

'This isn't about need—this is about his birthright. As my first born son he has the right to inherit the Lycander crown.'

'Even though he was born out of wedlock?'

It was the phrase her grandmother had used to describe Sunita's birth, to try and explain why her husband had thrown their pregnant daughter out.

'I know it is hard to understand in this day and age, Sunita, but in our family a mixed race child, born out of wedlock, was a stigma. It wasn't right, but it was how my husband felt.'

A feeling shared by others. Sunita could still feel the sting of the taunts her half-siblings had flung at her—nasty, insidious words that had clawed at her self-esteem.

Focus. Frederick watched her, his hazel eyes neutral and cool; he was in control and she quite clearly wasn't. Her thoughts raced round a playground of panic, visited the seesaw, spent time on the slide. Being born out of wedlock would have no impact on Amil's life; it was not a reason to get married.

She forced herself to concentrate on Frederick's answer to her question.

'It makes no odds as long as we legitimise him through marriage,' he said. 'Lycander's rules are complex, but clear on that front.'

Oh, Lord. What was she supposed to do? How could she make a decision like this without the use of a crystal ball? Her mother had believed the right course of action had been to hand Sunita over to her father.

'People can change, Suni,' her mother had said. She'd stroked Sunita's hair with a hand that had looked almost translucent, the effort of even that movement an evident strain. *'I have to believe that.'*

Sunita understood the uncharacteristic thread of sentimentality in her mother over those final weeks. Leela Baswani had wanted to die believing her daughter would be safe and happy, and so she had allowed herself to be conned again by the man who had already broken her heart. She'd allowed herself to believe that people could change.

Well, she'd been wrong. And so was this.

'This is impossible, Frederick. We can't spend the rest of our lives together.'

The very idea of spending a *week* with anyone made her skin prickle in affront—she could almost feel the manacles closing round her wrists. 'Maybe we should get married, legitimise Amil, and then get a divorce.'

Even as the words left her lips she knew how stupid they were.

'No. I want to give Amil a life with both his parents, and most importantly, if he is to rule Lycander, he needs to live in the palace, be brought up to understand his inheritance. And I need a wife—a true consort.'

This was becoming laughable. 'Really, I am *not* wife material—trust me on this.'

His broad shoulders lifted. 'But you *are* the mother of my child.'

Fabulous. 'So you'll make do with me because I come with a ready-made heir? And this whole marriage idea is because we are convenient?' The idea caused welcome anger—an emotion she could manage way better than panic.

'You don't care about Amil as a person—you only care about him as a commodity.'

'No!' Her words had clearly touched a nerve. 'I care about Amil because he is my son and I believe this is his right. I want him to grow up with two parents. And, believe me, this is hardly *convenient.* I intended to present my people with a wife and heir in a more conventional way.'

'Well, gee, thank you. That makes us feel *really* special.'

But she was the woman who had omitted to mention his son's existence—making her feel special would hardly be anywhere on his agenda.

His raised eyebrows indicated complete accord with her unspoken thought. 'There's no point in hypocrisy. If you expect me to go down on one knee, think again.'

'I don't expect anything—especially not a proposal. I don't want to marry you; I don't want to marry anyone.'

'I appreciate that. Until recently marriage has never exactly been high on my to-do list either. Back in the day I had a business to run and a party lifestyle to maintain. But circumstances have changed. For us both. *We* have Amil. *I* now have a country to run. I need a wife and I need an heir…to show the people of Lycander that I have changed. That I am responsible, that I offer stability, that I can put the principality's needs above my own.'

Sunita tried to equate this Frederick with the man she had known. '*Have* you changed?'

'Yes.' The syllable was bleak in its certainty, but despite its brevity it conveyed absolute conviction. 'You can choose to believe that or not, as you wish. But believe this: I need to get married.'

'Well, I don't. I prefer to be on my own.' She didn't want to be tied to anyone—she wanted to be independent and free to make decisions for herself and for Amil. 'Free.' *In control.*

'I understand that.' His jaw set in a hard line. 'But marriage is the only way to secure Amil his birthright and give him two parents one hundred per cent of the time.'

There was a strange undercurrent to his voice, and she realised just how important this must be to him. According to her research, his parents had split when he was three and his father had won sole custody. After that he'd had a series of stepmothers, none of whom had lasted more than a few years. So perhaps it was little wonder he wanted to give his son the kind of stable family he'd never had. For a moment, compassion for the little boy he had once been touched her and she forced herself to concentrate on the present.

'But it wouldn't be good for Amil to grow up and see his parents in an unhappy marriage.'

'Why assume it will be unhappy?'

'Because…' To her own annoyance, not a single reason sprang to mind that didn't sound stupid. Eventually she said, 'You can't expect me to sign up to a life sentence with a man I don't even know.'

'Fair enough. Then let's rectify that.'

He smiled—a smile of the toe-curling variety, like sunshine breaking through a grey cloudbank. And she couldn't help smiling back. But then the moment was gone and the stormy skies reappeared.

'Rectify it how?'

'Let's get to know each other. Bring Amil to Lycander and—'

'No! Once we are in Lycander I have no idea if we will be subject to Lycander law. Which, as far as I can gather, is *you*.'

The smile was a distant memory now, his face set in granite. 'You don't trust me?'

'I don't trust anyone.' After all, if you couldn't

trust your own father, who *could* you trust? His promise to her mother that he would look after Sunita, care for her as only a parent could, had turned out to be a bunch of empty, meaningless syllables.

'So we stay here.'

He raised his hands. 'Fair enough. But I can't be away for too long. I can stay in Mumbai for a few days or… Wait, I have a better idea.' The smile made a return. 'How about we go away for a few days? You and me. Away from the press and the politics and the spotlight.'

'You and me?' Panic and horror cartwheeled in her stomach.

'Yes. You and me. I'll put my money where my mouth is—you said you couldn't marry someone you didn't know, so here's the opportunity to spend time with me. Twenty-four-seven, with no distractions.'

'Be still my beating heart.'

Now his smile broadened and this time she was sure her hair curled.

'I *knew* you'd like the idea. Would your grandmother be happy to look after Amil?'

'If I agree to this, Amil comes with us.' A frown touched her brow and her eyes narrowed in suspicion. 'Surely you want to get to know your son?'

'Of course I do. But before we spend time together as a family, we need to know where we stand. I know he is only a baby, but I want him to have certainty and stability.'

The kind of certainty she guessed he'd never known. Again for an instant she wanted to reach out and offer comfort. *What to do? What to do?* In truth she didn't know. She should close this down now—but was that the right thing for Amil?

Frederick wanted to be a real part of his life, wanted to make him his heir. She couldn't in all conscience dismiss it out of hand. More than that, insane though it might be, there was a tiny part of her that didn't want to. That same tiny part that two years ago had wanted Frederick to ask her to stay, to sweep her into his arms and—

Cue mental eye-roll and a reality check. Fairy tales didn't exist. This was for Amil's sake.

'OK. Two days. I won't leave Amil longer than that.'

'Deal. Where do you want to go?'

Sunita thought for a moment. 'Goa.' That would keep it all in perspective—her parents had spent some time in Goa; they'd been happy there, but that hadn't led to a happily-ever-after in any sense.

'OK. Here's how it'll work. I'll have my people pick up Amil and your grandmother now, and bring them to the hotel. Once we make the announcement about Amil the press will converge. I want my son safe here, under royal security protection.'

She could feel the colour leech from her skin and saw that he had noticed it.

'I don't believe he is danger, but his position has changed. No matter what we decide, there will be more interest in him and his life from now on.'

She inhaled an audible breath. 'You're right. I'll call my grandmother and prepare her.'

Pulling her phone out of her bag, she rose and walked to the opposite end of the room.

Frederick watched as Sunita paced the width of the room as she talked, her voice low but animated, one hand gesturing as the conversation progressed.

It was impossible not to admire her fluidity of movement, her vibrancy. At least she hadn't blown the marriage idea out of the water. But he'd known she wouldn't do that—for Amil she had to consider it. What woman would deprive her son of a crown? Yet unease still tingled in his veins. Sunita might well be the one woman who would do exactly that.

Ironic, really—his chief advisor had a list of women who wanted to marry him, and he'd proposed to the one woman who didn't even want to audition for the part of bride.

No matter—he would convince her that this was the way forward. Whatever it took.

His conscience jabbed him. Really? *Whatever* it took? Maybe that was how his father had justified the custody battles.

Abruptly he turned away and, pulling his own phone out, set to work making arrangements.

He dropped his phone back in his pocket as she returned to the table. 'How did your grandmother take it?'

'With her trademark unflappable serenity. I think she suspected—she may even have recognised you earlier and put two and two together. She'll have Amil ready.' Her chin jutted out at a defiant angle. 'I've asked Sam and Miranda as well.'

She really didn't trust him. 'Do you really think I will take Amil from you by force?'

Silence greeted this and he exhaled heavily.

'If you can't trust my morality then at least trust my intelligence. I want you to marry me— kidnapping Amil would hardly help my cause. Or garner me positive publicity in Lycander. You hid Amil from me for two years. I have more reason to distrust *you* than vice versa.'

'Maybe it's best if neither of us trusts each other.'

She had a point.

'Works for me. Whilst we are away Amil will be in your grandmother's charge, with Sam and Miranda as your back-up. But they remain based in the hotel, and if they go anywhere one of my staff goes with them. Does that work for you?'

'Yes.'

'Good. Once they are safely here I'll announce it to the press. We'll leave for Goa tomorrow, after my visit to the school.'

'Whoa! Hold on.' One elegant hand rose in the air to stop him. 'This is a *joint* operation. So, first off, *I* want to make the announcement. And we are *not* mentioning marriage.'

She drummed her fingers on the table and he could see her mind whir. This had always been her forte—she'd used to play the press like a finely tuned instrument, and had always orchestrated publicity for maximum impact with impeccable timing.

'Prince Frederick and I are delighted to announce that fourteen months ago our son Amil was born. Obviously we have a great deal to discuss about the future, which we will be doing over the next few days. My press office will be in touch with details of a photo opportunity with the three of us tomorrow.'

'Photo opportunity?' Three of us...? The

words filled him with equal parts terror and anticipation.

'Yes. Better to arrange it than have them stalk us to try and get one. And I assume you want to spend time with Amil before we go?' She clocked his hesitation before he could mask it. 'Is there a problem?'

'No.' *Liar.*

Her eyes filled with doubt. He racked his brain and realised that in this case only the truth would suffice.

'I don't want to upset Amil or confuse him just before you leave him.'

He didn't want his son to believe on any level that it was his father's fault that he was losing his mother. Even for a few days.

For the first time since his proposal she smiled—a real, genuine smile—and he blinked at the warmth it conveyed. If he were fanciful, he'd swear it had heated his skin and his soul.

'You won't upset him. Truly. How about we take him to the Hanging Gardens? He loves it there—the press can take their photos and then we can take him for a walk.'

'Sounds great.'

But the warmth dissipated and left a cold sheen of panic in its wake. What if the meeting didn't go well? What if they couldn't connect?

Then he'd fake it. If he could close his emotions down—and he was a past master in the art—then surely the reverse would be true too. 'My school visit is planned for seven a.m., so if we schedule the press for midday that should work.'

'I'd like to come with you to the school. It's a cause I'd love to be involved in, and now... now I can.'

Her smile broadened and it occurred to him that, whilst he couldn't condone what she had done, hiding Amil had impacted on Sunita's life heavily. She'd lost her career, had to subdue her identity and become anonymous.

Sheesh. Get a grip. Any minute now he'd start to feel sorry for her.

The point now was that Sunita would be an asset to the charity.

His phone beeped and he read the message.

'Amil and your grandmother are in the hotel. So are Sam and Miranda. So let's go and face the press.'

And then he'd face the music. He had no doubt his chief advisor had set up a veritable orchestra.

CHAPTER SIX

'YOU'VE DONE *WHAT*?' Marcus Alrikson, hot off a private jet, scooted across the floor of the hotel suite. 'The whole existence of a secret baby is bad enough—but now you're telling me you have proposed marriage!' Marcus paused, pinched the bridge of his nose and inhaled deeply. 'Why?'

Frederick surveyed him from the depths of the leather sofa. 'Because I have a son, and I want my son to live with me *and* his mother. I realise that flies in the face of Lycander tradition, but there you have it. I want Amil to inherit his birthright. The only way to achieve both those goals is marriage.'

'If this marriage loses you the crown he won't have a birthright to inherit.'

'It won't.' Frederick imbued his voice with a certainty he was far from feeling—but he was damned if he would admit that to Marcus. 'This is the right thing to do and the people of Lycander will see that.'

'Perhaps…but that doesn't mean they will accept Sunita or Amil.'

'They will have no reason not to. Sunita has proved herself to be an exemplary mother. And she will be an exemplary princess.'

Marcus shook his head. 'She is a supermodel with a reputation as a party girl. You have no idea what she may or may not do—she would never have made my list in a million years. She is as far from Kaitlin Derwent as the moon is from Jupiter.'

'And look what happened with Lady Kaitlin. Plus, don't you think you're being a little hypocritical? What about *my* reputation?'

'You have spent two years showing that you have changed. The reforms you are undertaking for Lycander are what the people want. You may have been a playboy with a party lifestyle, but you also founded a global business—Freddy Petrelli's Olive Oil is on supermarket shelves worldwide. At least you partied on your own dime.'

'So did Sunita. And her party days were over by the time we met.'

'Sunita has spent two years hiding your son from you,' Marcus retorted. 'There is nothing to suggest she will be good for Lycander and plenty to suggest she will plunge the monarchy straight back into scandal. She could run off with Amil, file for divorce before the honeymoon is over…'

'She won't.'

He couldn't know that, though—not really. He'd known Sunita for a couple of weeks two years ago. Doubt stepped in but he kicked it out even as he acknowledged the sceptical rise of his chief advisor's eyebrow. 'Or at least it's a risk I am willing to take.'

'It is too big a risk. The women on my list are open to the idea of an arranged marriage—they have been brought up to understand the rules. Dammit, Frederick, we *had* this discussion. We agreed that it was important for the Lycander bride to be totally unlike your father's later choices and more in line with his first wife.'

Axel's mother, Princess Michaela, a princess in her own right, had been a good woman.

'We did. But circumstances have changed.'

'Doesn't matter. You plan to present your people with a bride who may well cause a scandal broth of divorce and custody battles.'

'I have no choice—none of this is Amil's fault.'

'I am not suggesting you turn your back on Amil. Provide for him. See him regularly. But do not marry his mother.'

'No.' It was as simple as that. 'I will be a real father to Amil and this is the way forward. I'm doing this, Marcus—with or without your help.'

Silence reigned and then Marcus exhaled a long sigh and sank into the seat opposite. 'As you wish.'

Sunita surveyed her reflection in the mirror, relieved that there was no evidence of the tumult that raged in her brain. Frederick…discovery…Amil…Crown Prince…marriage…Goa…disaster.

There was potential disaster on all fronts—the thought of marriage was surreal, the enormity of the decision she needed to make made her head whirl and the idea of two days in Goa with Frederick made her tummy loop the loop.

A tentative knock on the door heralded the arrival of Eric, the Lycander staff member who had been dispatched to her apartment the previous night.

'Good morning, Eric, and thank you again for getting my things. I really appreciate it.'

'You're very welcome, ma'am. The Prince is ready.'

She followed Eric through the opulence of the hotel, with its gold and white theme, along plush carpet and past gilded walls, through the marble lobby, past luscious plants and spectacular flower arrangements and outside to the limo. There Frederick awaited her, leant against the hood of the car, dressed casually in jeans and a

T-shirt, his blond hair still a touch spiked with damp, as if he'd grabbed a shower on the run.

'A limo? Isn't that a touch ostentatious?'

The flippant comment made to mask her catch of breath, the thump of her heart. 'I promised the children a limo after my last visit—they were most disappointed when I turned up in a taxi.'

He held the door for her and she slid inside, the air-conditioned interior a welcome relief against the humidity, with its suggestion of imminent monsoon rain.

'They are amazing kids—they make you feel…humble.'

Sunita nodded. 'I read up on the charity last night. The whole set-up sounds awesome and its achievements are phenomenal. I love the simplicity of the idea—using open spaces as classrooms—and I admire the dedication of the volunteers. I'll do all I can to raise the profile and raise funds. Today and in the future.'

'Thank you. Axel helped set up the charity and donated huge sums after someone wrote to him with the idea and it caught his imagination. I wish…'

'You wish what?' The wistfulness in his voice touched her.

'I wish he'd told me about it.'

'People don't always like to talk about their charitable activity.' She frowned. 'But in this case surely he must have been pretty public about it, because his profile would have raised awareness.'

'It didn't work like that. My father was unpredictable about certain issues—he may not have approved of Axel's involvement. So Axel kept it low-key. Anonymous, in fact. I only found out after his death because someone from the charity wrote with their condolences and their thanks for all he had done. I decided to take over and make it a more high-profile role.'

'Didn't your father mind?'

Frederick shrugged. 'I don't know. He and I weren't close.' His tone forbade further questions. 'Anyway, in the past two years the number of schools has increased three-fold and I've hired an excellent administrator—she isn't a volunteer, because she can't afford to be, but she is worth every penny. The schools are makeshift, but that has saved money and I think it makes them more accessible.'

His face was lit with enthusiasm and there was no doubting his sincerity. Any reservations she'd harboured that this was simply a publicity stunt designed to show that Frederick had a charitable side began to fall away.

This continued when they arrived at the school and a veritable flock of children hurtled towards the car.

He exhibited patience, good humour and common sense; he allowed them to feel and touch the car, and then promised they could examine the interior after their lessons—as long as their teacher agreed.

A smiling woman dressed in a forest-green and blue *salwar kameez* came forward and within minutes children of differing ages and sizes were seated in the pavilion area and the lesson commenced.

Sunita marvelled at the children's concentration and the delight they exuded—despite the open-air arena, and all the distractions on offer, they were absorbed in their tasks, clearly revelling in the opportunity to learn.

'Would you like to go and look at their work?' the teacher offered, and soon Sunita was seated next to a group of chattering children, all of whom thrust their notebooks towards her, emanating so much pride in their achievements that flipped her heart.

She glanced at Frederick and her heart did another turn. Standing against a backdrop of palm trees and lush monsoon greenery, he was performing a series of magic tricks that held the children spellbound. He produced coins from ears and cards from thin air, bringing gasps of wonder and giggles of joy.

Finally, after the promised exploration of the

limo, the children dispersed—many of them off to work—and after a long conversation with the teacher Frederick and Sunita returned to the car for their journey back to the hotel.

'Next time I'll take Amil,' Sunita said. 'I want him to meet those kids, to grow up with an understanding of the real world.'

'Agreed.'

The word reminded Sunita that from now on Frederick would have a say in her parenting decisions, but right now that didn't seem to matter. This was a topic they agreed on.

'There's such a lesson to be learnt there—those children want to learn, and it doesn't matter to them if they have computers or science labs or technology. They find joy in learning, and that's awesome as well as humbling.'

'*All* of this is humbling.'

He turned to look out of the window, gesturing to the crowded Mumbai streets, and Sunita understood what that movement of his hand had encompassed—the poverty that was rife, embodied by the beggars who surged to the limo windows whenever the car slowed, hands outstretched, entreaty on their faces. But it was more than that – you could see the spectrum of humanity, so many individuals each and every one with their own dreams and worries.

'You really care. This isn't all a publicity stunt…part of your new image.'

'This is about a continuation of Axel's work—no more, no less. Don't paint me as a good person, Sunita. If it weren't for Axel I would never have given this so much as a thought.'

The harshness of his voice shocked her, jolted her backwards on the seat with its intensity. 'Perhaps, but you were hardly duty-bound to take over—or to come out here and interact with those children like that.' She couldn't help it. 'Axel didn't do that, did he?'

'Axel *couldn't* do that—he needed to be the heir my father wanted him to be.'

With that he pulled his phone out of his pocket in a clear indication that the subject was well and truly closed.

Sunita frowned, fighting the urge to remove the phone from his grasp and resume their conversation, to make him see that he was wrong—in this instance he *was* a good person.

Back off, Sunita.

Right now she needed to remain focused on whether or not she wanted to marry this man—and what the consequences of her decision would be for Amil. And in that vein she needed to look ahead to the photo call, which meant an assessment of the recent press coverage. So she pulled her own phone out of her pocket.

A few minutes later he returned the mobile to his pocket.

'OK. We'll fly to Goa late afternoon, after the photo call and the trip to Hanging Gardens. As you requested I've sorted out a room near your suite for Sam and Miranda.'

'Thank you. I appreciate that.'

Goa! Sudden panic streamed through her and she pushed it down. She was contemplating *marriage* to Frederick, for goodness' sake—so panic over a mere two days was foolish, to say the least. She needed to focus on Amil.

She glanced across at Frederick, wondering how he must feel about taking Amil out. Perhaps she should ask, but the question would simply serve as a reminder of the fact that he had missed out on the first fourteen months of his son's life.

So instead she faced forward and maintained silence until the limo pulled up outside the hotel.

Frederick stood outside the hotel bedroom door. His heart pounded in his chest with a potent mix of emotions—nervousness, anticipation and an odd sense of rightness. In two minutes he would meet Amil. Properly. Terror added itself to the mix, and before he could turn tail and flee he raised his hand and knocked.

Sunita opened the door, Amil in her arms, and he froze. He didn't care that he was standing in the corridor in full sight of any curious passers-by. All he could do was gaze at his son. *His son*.

Wonder entered his soul as his eyes roved over his features and awe filled him. *His son*. The words overwhelmed and terrified him in equal measure, causing a strange inability to reach out and hold the little boy. His emotions paralysed him, iced his limbs into immobility, stopped his brain, brought the world into slow motion.

Determination that he would not let Amil down fought with the bone-deep knowledge that of course he would. He wasn't equipped for this—didn't have the foundations to know how to be a parent, how not to disappoint.

But he would do all he could. He could give this boy his name, his principality, and perhaps over time he would work out how to show his love.

Amil gazed back at him with solemn hazel eyes and again panic threatened—enough that he wrenched his gaze away.

'You OK?' Sunita's soft voice pulled him into focus and he saw understanding in her eyes, and perhaps even the hint of a tear at the edge of her impossibly long eyelashes.

'I'm fine.'

Get a grip.

He had no wish to feature as an object of compassion. So he kept his gaze on Sunita, absorbed her vibrant beauty, observed her change of outfit from casual jeans and T-shirt to a leaf-print black and white dress cinched at the waist with a wide red belt. Strappy sandals completed the ensemble.

'Babababababa!' Amil vouchsafed, and a well of emotion surged anew.

'Do you want to hold him?'

'No!' *Think.* 'I don't want to spook him—especially just before the photo call.'

It wasn't a bad cover-up, but possibly not good enough to allay the doubts that dawned in her eyes.

'You won't. He's fairly sociable. Though obviously he doesn't really meet that many strange me—' She broke off. 'I'm sorry. Of all the stupid things to say that took the cake, the biscuit and the whole damn patisserie.'

'It's OK. I am a stranger to Amil—that's why I don't want to spook him.'

His gaze returned to the baby, who was watching him, his eyes wide open, one chubby hand clutching a tendril of Sunita's hair.

'We need to go.'

'I know. But first I have a couple of questions about the press conference and the Kaitlin question.'

Frederick frowned. 'What question would that be?'

'A couple of reporters said, and I quote, that you are "broken-hearted" and that perhaps I can mend the chasm. Others have suggested you would welcome a dalliance with an old flame as a gesture, to show Lady Kaitlin you are over her.'

'I still don't understand what your question is.'

'Two questions. *Are* you heartbroken? *Are* you over her?'

'No and yes. I need to get married for Lycander. My heart is not involved. Kaitlin understood that—our relationship was an alliance. When that alliance became impossible we ended our relationship. Since then she has met someone else and I wish her well.'

Sunita's expression held a kind of shocked curiosity. 'That's *it*? You were with her for *months*. You must have felt *something* for her.'

Momentary doubt touched him and then he shrugged. 'Of course I did. I thought that she would be an excellent asset to Lycander.'

Kaitlin's diplomatic connections had been exemplary, as had her aristocratic background. She'd had a complete understanding of the role of consort and had been as uninterested in love as he was.

'I was disappointed when it didn't work out.'

'Yes. I see that it must have been tough for you to have the deal break down.' Sarcasm rang out from the spurious sympathy.

'It was—but only because it had an adverse impact on my position as ruler.'

And that was all that mattered. His goal was to rule Lycander as his brother would have wished, to achieve what Axel would have achieved. Whatever it took.

'So all you need to know about Kaitlin is that she is in the past. My heart is intact.' He glanced at his watch. 'And now we really need to go.'

A pause and then she nodded. 'OK. This is our chance to change the mixed reaction into a positive one. An opportunity to turn the tide in our favour.'

'You sound confident that you can do that.'

'Yup. I'm not a fan of bad publicity. Watch and learn.'

One photo call later and Frederick was looking at Sunita in reluctant admiration. He had to hand it to her. By the end of the hour she had had even the most hostile reporter eating out of her hand. Somehow she had mixed a suggestion of regret over her actions with the implicit belief that it had been the only option at the time. In addition, she had managed to make it clear that whilst two years ago Frederick had been a shallow party prince, now he had morphed into a different and better man, a worthy ruler of Lycander.

No doubt Marcus had been applauding as he watched.

Hell, even *he* had almost believed it. *Almost.*

'You did a great job. And I appreciate that you included me in your spin.'

'It wasn't spin. Everything I said about you was true—you *have* worked incredibly hard these past two years, you *have* instigated all the changes I outlined, and you *do* have Lycander's future at heart.'

The words washed over him like cold, dirty water—if the people of Lycander knew where the blame for Axel's death lay they would repudiate him without compunction, and they would be right to do so. But he didn't want these thoughts today—not on his first outing with Amil.

He glanced down at Amil, secure now in his buggy, dressed in a jaunty striped top and dungarees, a sun hat perched on his head, a toy cat clasped firmly in one hand.

'Amamamamam…ma.' Chubby legs kicked and he wriggled in a clear instruction for them to move on.

Sunita smiled down at her son. 'I think he wants to get going—he wants to see all the animal hedges. They seem to utterly fascinate him.'

As they wandered through the lush gardens that abounded with shades of green tranquillity seemed to be carried on the breeze that came from the Arabian Sea, and for a moment it was almost possible to pretend they were an ordinary family out for the day.

Sunita came to a halt near a topiary hedge, one of many clipped into the shape of animals. 'For some reason this is his favourite—I can't work out why.'

Frederick studied it. 'I'm not sure I can even work out what it is. I spotted the giraffe and the elephant and the ox-drawn cart, but this one flummoxes me.'

Sunita gave a sudden gurgle of laughter. 'I know what Amil thinks it is. Amil, sweetheart, tell Mu— Tell us what the animal does.'

The little boy beamed and made a *'raaaah'* noise.

Frederick felt his heart turn over in his chest. Without thought he hunkered down next to Amil and clapped. 'Clever boy. The tiger goes *"rah"*.'

'Raaaah!' Amil agreed.

And here it came again—the paralysis, the fear that he would mess this up. He'd never managed any other relationship with even a sliver of success. Why would this be different?

Rising to his feet, he gestured around the garden. 'This is a beautiful place.'

'I used to come here as a child,' Sunita said. 'It's one of my earliest memories. I loved the flower clock.'

She pressed her lips together, as if she regretted the words, and Frederick frowned. Her publicity blurb skated over her childhood, chose to focus instead on her life after she'd embarked on her career. Almost as if she had written her early years out of her life history...

'Come on,' she said hurriedly. 'This morning isn't about my childhood. It's about Amil's— let's go to the Old Woman's Shoe.'

Five minutes later Frederick stared at the shoe—actually an enormous replica of a boot. As landmarks went, it seemed somewhat bizarre— especially when the words of the nursery rhyme filtered back to him.

There was an old woman who lived in a shoe.
She had so many children she didn't know what to do.
She gave them broth without any bread,
Then whipped them all soundly and sent them to bed.

'Isn't this a slightly odd thing to put in a children's playground?'

'Yes. But I loved it—I used to climb it and it made me feel lucky. It was a way to count my blessings. At least I didn't live with a horrible old woman who starved me and beat me!'

At least. There had been a wealth of memory in those syllables, and for a daft moment he had the urge to put his arm around her and pull her into the comfort of a hug.

As if realising she had given away more than she had wanted, she hastened on. 'Anyway, I looked up the rhyme recently and it turns out it probably has political rather than literal connotations. But enough talk. This is about you and Amil. Do you want to take Amil into the shoe? I'll wait here with the buggy.'

The suggestion came out of nowhere, ambushed him, and once again his body froze into immobility even as his brain turned him into a gibbering wreck.

'I think that may be a little bit much for him. He barely knows me.' *Think*. 'We haven't even explained to him who I am.'

The accusation in his own voice surprised him—and he knew it masked a hurt he didn't want her to see. Because it exposed a weakness he didn't want her to know. *'Never show weakness, my son.'* The one piece of paternal advice he agreed with. *'Show weakness and you lose.'* Just as all his stepmothers had lost. Their weakness had been their love for their children—a weakness Alphonse had exploited.

Heat touched the angle of her cheekbones as she acknowledged the truth of his words. 'I know. I'm not sure what you want to do. I don't know what you want him to call you. Dad? Daddy? Papa?'

In truth he didn't know either, and that increased his panic. Sunita stepped towards him, and the compassion in her eyes added fuel to the panic-induced anger.

'But remember, he is only fourteen months old—I don't think he understands the concept of having a dad.'

The words were a stark reminder of her deception.

'Amil doesn't understand or *you* don't?'

The harshness of his voice propelled her backwards, and he was glad of it when he saw the compassion vanish from her expression.

'Both of us. Give me a break, Frederick. Until yesterday it was just Amil and me. Now here you are, and you want to marry me and make Amil the Crown Prince. It's a lot to take in.'

For an instant he empathised, heard the catch in her voice under the anger. But this was no time for empathy or sympathy. Now all that mattered was the knowledge of what was at stake.

'Then take it in fast, Sunita. You chose to hide Amil from me and now you need to deal with the consequences of that decision. Most people wouldn't think they were so bad. *I* am the one who has missed out on the first fourteen months of my son's life. *My son*. I am Amil's father and *you* need to deal with it.'

There was silence, broken only by the sound of Amil grizzling, his eyes wide and anxious as he looked up at Sunita.

Oh, hell. Guilt twisted his chest. What was *wrong* with him? This was his first outing with Amil and he'd allowed it to come to this. Shades of his own father, indeed.

He squatted down beside the baby. 'I'm sorry, Amil. Daddy's sorry.' Standing up, he gestured to the Old Woman's Shoe. 'You take him up. I'll wait here with the buggy. I've upset him enough—I don't want to compound my error.'

Sunita hesitated, but then Amil's grizzling turned to tears and she nodded assent.

'OK.' Leaning down, she unbuckled Amil and took him out. 'Come on, sweetheart. Let's try some walking.'

Frederick watched their progress and determination solidified inside him. He might be messing this up big-time, but he would not concede defeat. At the very least he would give his son the chance to be a prince. Their outing to the Hanging Gardens might be a disaster, but going to Goa wouldn't be.

By the end of their time there Sunita would agree to marry him.

CHAPTER SEVEN

SUNITA LOOKED ACROSS the expanse of the royal jet to where Frederick sat. There was no trace of the man she'd glimpsed mere hours ago in the Hanging Gardens—a man who had exhibited a depth of pain and frustration that had made her think long and hard.

Another glance—he still looked cool, regal and remote, and she couldn't read any emotion or discern what thoughts might be in his mind. Which would make what she had to say all the more difficult.

For a moment she nearly turned craven. *No.* This was the right thing to do and she would do it.

'Frederick?'

'Sunita.'

'Can we talk?'

'Of course.' He pushed his netbook across the table, rose and crossed to sit in the luxurious leather seat next to hers. 'Shoot.'

'I've thought about what you said earlier. About me having to accept that you are Amil's father.'

He raised a hand. 'It doesn't matter. I shouldn't have said what I did.'

'It *does* matter. I don't see how we can even consider a future together until we resolve our past. So I want to say I'm sorry.'

She twisted her hands together on her lap, re-calling Frederick's expression when he'd looked at Amil as if his son was the most precious being in the universe.

'I'm sorry you missed out on Amil's first months.'

However justified her decision, Frederick could never have that time back—would never be able to hold his newborn son in his arms, see his first smile, run his finger over his gum to reveal that first tooth.

'I'm sorry.'

'OK.'

'But it's *not* OK, is it?'

'No.' He closed his eyes, then reopened them. 'No. It isn't OK that you hid my son's existence from me.'

'I couldn't take the risk.'

'Yes. You could have. You *chose* not to.'

Rationalisations lined up in her vocal cords but she uttered none of them. Bottom line—he was right. Her choice had meant Frederick had missed out on something infinitely precious.

'Yes, I did. And all the reasons I gave you earlier were true. But it's more than that.'

She inhaled deeply. She had no wish to confide this to him—she wasn't even sure she wanted to acknowledge it herself. But there it was again—the memory of the way Frederick had looked at Amil, the fact that he wanted to be part of his son's life and wanted to create a stable family unit. He deserved a true explanation.

'I thought history was repeating itself. I thought you would be like...' Her voice trailed off, her brain wishing it could reverse track and pull the words back.

'Like who?'

The gentleness of his voice surprised her—gave her the momentum to carry on.

'Like my father. He was a Londoner, on holiday in India with a group of friends when he met my mother. They fell in love—or so she believed. She fell pregnant and she *did* choose to tell him, and all she could see was a tornado of dust as he disappeared. Straight back, roadrunner-style, to his fiancée in London.'

Even now the enormity of her father's selfishness had the power to stun her—he *must* have understood the repercussions. They would have been complex enough in any culture, but in India there had been added layers of complication that transcended even betrayal and heartbreak.

Understanding showed in the expression on Frederick's face. 'That must have been tough for your mother.'

'Yes. It was. It changed the entire trajectory of her life. Her family was horrified and threw her out—she was only nineteen, and she had to fend for herself in a society which by and large had condemned her. And a lot of that is down to my father and his rejection of her—and me. I know we were in different circumstances—you didn't lie to me—but I knew you didn't want children. I didn't want to hear you say the same words my own father had—I didn't want Amil to feel the sense of rejection I did.'

Sunita forced herself to hold his gaze, to keep her tone level. This verged on the excruciating—touchy-feely confidences were not her bag at all.

'It seemed better, easier, less painful, to bring Amil up on my own. I figured what *he* didn't know and *you* didn't know wouldn't hurt anyone.'

There was a silence, and then he reached out, touching her forearm lightly. 'I'm sorry for what happened to your mother and to you. I promise you—I will never reject Amil.'

There could be no doubt as to the sincerity in his voice, and in the here and now she believed he meant every word. But she knew that good intentions did not always turn into actions. Her father must have once believed the empty promises he'd made to make up for his past, to be a good parent.

'It will not happen,' he repeated, as if he sensed her doubts. 'And now let's put the past behind us. I wish you had told me about Amil earlier, but I do understand why you made the choices you did. I believe now that we need to move forward, put the past behind us and focus on our present and our future. Deal?'

He held out his hand and Sunita looked down at it. So perfect—strong, masculine, capable… Capable of the gentlest of caresses, capable of…

Close it down, Sunita.

Too late—images scrambled her mind and for a moment she was unable to help herself. She closed her eyes, let the sensation dance over her skin. But it was more than desire—she knew that this deal signified understanding and forgiveness, and that made her head whirl as well.

Then she opened her eyes and reached out, clasped his hand and worked to still the beat of her heart. 'Deal,' she said. The syllable emerged with way too much violence, and she dropped his hand as if it were burning her. Which in a sense it was.

She looked down, then sneaked a look up at

him—had he seen her reaction? Of course he had. It didn't take a forensic degree to know that. Embarrassment flushed her skin even as she couldn't help but wonder if this stupid physical reaction was a mutual one.

Her gaze met his and against all odds her pulse quickened further. His hazel eyes had darkened, the heat in them so intense her skin sizzled as her hormones cartwheeled.

Nothing else mattered except this.

Her lips parted as he rose, and his eyes never once left hers as he held out a hand. Without thought she put her hand in his, and he tugged her up so they stood mere centimetres apart.

Oh, so gently, but with a firmness that neither expected nor brooked denial, his hands encircled her waist and pulled her body flush against his. The feel of him, of the hard, muscular wall of his chest, made her gasp, and she looped her arms round his neck, accidentally brushing the soft skin on his nape.

An oath dropped from his lips and then those self-same lips touched hers and she was lost.

The kiss oh-so-familiar and yet so much more than before; the tang of coffee and the hint of strawberry jam, the sheer rollick of sensation that coursed her blood, made her feel alive and made her want more. He deepened the kiss and she pressed against him, caught in this moment that felt so damn right.

Stop. What the hell was she doing?

She wrenched out of his arms so hard she nearly tumbled over, putting a hand out to steady herself against the back of the chair.

For a moment silence reigned, broken only by the sound of their jagged breathing. Sunita tried to force herself to think through the fog of desire that refused to disperse. She couldn't let herself succumb to him again—she *couldn't*. Two years ago she'd lost her self-respect—now she could lose even more than that. Her attraction was a weakness he could play on—something that might cloud her judgement when she needed it most.

'I'm sorry. That was stupid.'

He ran a hand down his face, almost as if to wipe away all emotion, all desire, and when he met her gaze his expression was neutralised. 'No need to be sorry. That was a *good* thing.'

'How do you figure that out?'

'Because it proves we have physical compatibility. That's important in a marriage.'

His words acted like the equivalent of a bucket of ice-cold water and she slammed her hands on her hips. 'So that kiss was a deliberate ploy? A way to make the marriage more acceptable to me?'

'It wasn't a deliberate ploy, but it wasn't a mistake either. Mutual attraction is a benefit in a marriage. A bonus to our alliance.'

A benefit. A bonus. Any minute now he'd tell her there was some tax advantage to it too.

Sheer outrage threatened at his use of their attraction as a calculated move to persuade her. More fool her for believing he had been as caught up and carried away as she had. This *was* the Playboy Prince, after all.

'Well, I'll bear that in mind, but given that you have found "physical compatibility" with hundreds of women, I'm not sure it counts for much. Now, if you'll excuse me, I'll just go and freshen up.'

Frederick resisted the urge to put his head in his hands and groan. Then he considered the alternative option of kicking himself around the private jet.

Kissing Sunita had not been on the agenda—but somehow her beauty, her vulnerability, her honesty had overwhelmed him, and what he had meant to offer as comfort had turned into the type of kiss that still seared his memory, still had his body in thrall.

Dammit. He would *not* let physical attraction control him as it had his father—that way led to stupid decisions, poor judgement calls and people getting hurt. Yet during that kiss his judgement could have parachuted off the plane and he wouldn't have given a damn.

Then, to compound his original stupidity, he had morphed into a pompous ass. Words had flowed from his tongue as he'd fought the urge to pull her straight back into his arms and resume proceedings. What an idiot. And then there had been her reference to his past. The truth was, even back then Sunita had been different from his so-called 'hundreds of women'.

He looked up as she returned to the room, her brown eyes cold, her expression implacable as she headed back to her chair and reached down into her bag for a book.

Hell. Now what? This was not going to plan and he didn't know how to retrieve it. Did not have a clue. He was so far out of his comfort zone he'd need a satnav and a compass to find his way back.

'Sunita?'

'Yes.'

'That kiss…'

'I think we've said all that needs to be said about it. As far as I am concerned, I plan to erase it from my memory banks.'

'Fine. But before you do that I want to clarify something. You mentioned my "hundreds of women"—for starters, that is an exaggeration.

Yes, I partied hard and, yes, there were women, but not as many as the press made out. But, any which-way, those days are over and they have been for a long time. I was never unfaithful to any woman and I plan on a monogamous marriage.'

Clearly his default setting today was 'pompous ass', so he might as well run with it.

'So you'd be faithful for the duration. For decades, if necessary?'

The scepticism in her tone rankled.

'I am always faithful.'

'But your relationships have only lasted a few weeks at a time—that's hardly much of a test. Variety was the spice of your life.'

'Very poetic. Let's take it further, then—I believe it's possible to have variety *and* plenty of spice with one woman.'

'Then why didn't you ever try it before?'

Damn. Poetic *and* sharp.

'Because short-term suited me—I didn't want physical attraction to develop into any expectations of marriage or love. I never offered more than I could give and the same goes now. I can offer marriage and fidelity, but not love.'

'I still don't buy it. Most people are faithful *because* of love—if you don't believe in love what would motivate you to be faithful?'

'I will not repeat my father's mistakes. He

went through women like a man with a cold does tissues. Any beautiful woman—he thought it was his right to have her, whether he was already in a relationship or not, and it led to a whole lot of strife and angst. So I will not plunge Lycander into scandal and I will not hurt my children or humiliate my wife. That is nothing to do with love—it is to do with respect for my country and my family.'

'OK.'

Sympathy warmed her eyes and the moment suddenly felt too weighted, too heavy, and he cleared his throat. 'I thought you might want to know more about Lycander—after all, it will be your new home and your country.'

'I'd like that. I do remember some of what you told me two years ago. Rolling countryside, where you can walk and smell the scents of honeysuckle and almost taste the olives that you grow. You made the olive groves come to life.' She hesitated, and then asked, 'What happened to your business deal? The one you hoped would go through two years ago?'

Her words caused him to pause. Sunita had been one of the very few people he'd spoken to about his dreams. Ever since he was young he'd been focused on breaking free of his father's money—sick and tired of the constant reminders that he relied on his father's coffers for his food, his clothes, the roof over his head.

Then, at twenty-one, he'd come into the inheritance of a run-down, abandoned olive grove. And as he'd walked around it had been as if the soil itself had imparted something to him, as if the very air was laden with memories of past glories, of trees laden with plump lush olives, the sound and whir of a ghostly olive press.

That was where it had all started, and over the years he'd built an immensely profitable business. Two years before he'd been in the midst of a buy-out—he'd succeeded, and taken his company to the next echelon. That had been the deal he'd been celebrating—the reason he'd handed over the state function to Axel, the reason Axel had died.

Guilt and grief prodded him and he saw Sunita frown. *Focus.* 'The deal went through.'

'So who runs your business now?'

'A board of directors and my second-in-command—I have very little to do with it any more.'

'That must be hard.'

'That's how it is. Lycander needs my attention, and its people need to see that they come first. The principality isn't huge, but we have beaches, we have vineyards, we have olive groves. I know I'm biased, but our olives are the best in the world—they have bite...their taste lingers on your tongue—and the olive oil we produce is in a class of its own. As for our grapes—I believe the wine we produce rivals that of France and Spain. Lycander has the potential to be a prosperous land, but right now it is a vessel of past glories. My father increased taxes, lowered the minimum wage—did all he could to increase the money in the royal coffers without a care for the effect.'

'But couldn't anyone stop him?'

'No. In Lycander, the ruler's word is law—he has the final say on the governing of the land. Of course there are elected advisors, but they have no legislative power and the monarch can disregard their advice. So effectively everything hinges on having a ruler who genuinely cares about Lycander and its people.'

'That sounds like a whole heap of responsibility. For you. *And* to wish upon Amil.'

'It is, but I think it needs to be seen in context. In the past, when everything worked, it was easier—right now it is harder. But I will make sure I set things to rights. I know what needs to be done. I will make the laws fair, I will reduce taxation rates and I'll stop tax evasion. I want the divide between the wealthy and the poor to be bridged. I—'

He broke off at her expression.

'You can pick your jaw up from the ground.'

She raised her hand in admission. 'OK. Busted. I *am* surprised. Two years ago you were passionate about your business, but you didn't mention politics or social beliefs. Now your enthusiasm, your beliefs, are palpable.'

The all too familiar push and pull of guilt tugged within him.

'This isn't about my enthusiasm or my beliefs. It is about Axel—it's about fulfilling a promise. The people and the country suffered under my father's rule. The real reason there was no rebellion was that they knew one day Axel would succeed him, and that kept the unrest at bay. Axel had a vision—one that I *will* make happen.'

That had been the promise he'd made in his very first speech and he would fulfil it.

'What about *your* vison? The way you speak of Lycander—I can hear your pride in it.'

'I never had a vision for Lycander. I had a work hard, play hard lifestyle.'

'But you've changed?'

'Yes, I have.'

But the cost of that had been his brother's life.

Her frown deepened. She leant forward and he could smell her exotic scent with its overtone of papaya, could see the tiny birthmark on the angle of her cheekbone.

'I know you will be a good ruler. Whether

you rule because it is your duty or because your heart is in it.'

There was silence. She was close. Way too close. And he had had a sudden desire to tell her the truth about his ascent to the throne—a desire mixed with the longing to tug her back into his arms and damn common sense and practicality.

Neither could happen, so he rose to his feet and looked down at her.

'Thank you. But the point I was trying to make is that I will ensure the principality Amil inherits will be a *good* place, with a strong economic foundation. Of course he will still have much responsibility, but I hope it will not be a burden.'

'What if he doesn't want the job? What if he has other ambitions, other aspirations?'

'I would never force him to take the crown. He could abdicate.' He met her gaze. 'Provided we have more children.'

'More children?' she echoed.

'Yes. I would like more children in order to secure the succession.' After all, there was no hope of his brothers ever having anything to do with Lycander. 'To take the pressure off Amil.'

'Is that the only reason?'

'For now. I haven't really got my head around having Amil yet.'

Right now he was terrified about his ability

to parent *one* child—it wasn't the moment for a rose-tinted image of a functional, happy group of siblings.

'Do *you* want more kids?'

Sunita hesitated. 'I don't know...' A small smile tugged her lips upwards. 'I haven't really got *my* head around it all yet either. Until yesterday it was just me and Amil. My happiest memories are of my mother and me—just us. After—'

She broke off, looked away and then back at him, and he wondered what she had been about to say.

'Anyway,' she resumed, 'I'm not sure that the whole "happy family" scenario always works. Are you close to your other brothers?'

'No.'

His half-siblings... Stefan, who loathed all things Lycander, had left the principality as soon as he'd reached eighteen and hadn't returned. The twins, Emerson and Barrett, still only twenty, had left Lycander only days after their father's death and hadn't returned.

There was a definite pattern there, and it wasn't woven with closeness. The way they had grown up had made that an impossibility—their father had revelled in pitting brother against brother in a constant circus of competition and rivalry, and in the end Frederick had retired from the field, isolated himself and concentrated on his own life.

'But that was down to our upbringing. I hope that our children would do better.'

Perhaps it was a fruitless hope—there was every chance he would prove to be as useless a parent as his own parents had been, in which case perhaps a large family was a foolish idea.

But now wasn't the moment to dwell on it.

Relief touched him as the pilot announced their descent to Goa before Sunita could pursue the conversation further.

CHAPTER EIGHT

SUNITA'S EYES STRETCHED so wide she wondered if
her eyeballs would actually pop out of her head.

'This is incredible.'

In truth it was beyond incredible—and she
hadn't even seen the inside of the villa yet.

The drive itself had been unexpected—their
chauffeur-driven car had traversed remarkably
peaceful roads until they'd reached an idyllic vil-
lage seemingly untouched by tourism. Winding
lanes had displayed a number of villas draped
with greenery, and now they had arrived at
Sangwan Villa.

The Portuguese-built, newly renovated building was nestled amidst verdant grounds where
teak and jackfruit trees thrived, giving the air
an evocative smell of leather with a hint of pineapple.

Her gaze rested on the structure itself. With
its pillared verandas and high roof it looked like
a vision out of a fairy tale.

The thought jolted her. She needed to remember that fairy tales were exactly that—tales, fiction. And most fairy tales had a dark side, a grim under-story, and the myths they were built on didn't have any happily-ever-afters.

'How on earth did you get it at such short notice?'

'It was closed for maintenance—I made it worth the owners' while to postpone the work.'

A woman walked towards them, a smile on her face, her white and green sari very much in keeping with the verdant backdrop.

'Your Highness. Welcome. I am Deepali and I will be looking after you during your stay. Your staff have been settled in and your suites are ready, if you will follow me. I will show you your rooms and then I thought you may wish to have an evening drink by the pool before dinner. There are menus in your rooms—just call through when you are ready.'

'That sounds wonderful,' Sunita said. 'And thank you so much for making this available at such short notice.'

Minutes later she was looking around a sumptuous suite. 'It's beautiful…'

But it was more than that—it was quirky and cosy, with its warm aura countered by the cool of the tiled floor. The sitting area boasted comfy overstuffed armchairs, where she could imagine curling up with a book and a cup of coffee, or simply gazing at the courtyard outside, resplendent with shrubbery. Two steps led down to the bedroom, where a luxurious wooden bed sprawled against decadent red walls.

Her suitcases had been deposited by a large lacquered wardrobe and she opened one, needing the confidence fresh clothes would give her. A floaty dress with a vivid bird print gave her instant cheer, and as she made her way out to the courtyard she allowed herself to revel in the sound of kingfishers and the sight and scent of the opulent lilies in the ornate pond.

Frederick sat on a recliner chair, a frosted beer bottle on the small table behind him and his blond head slightly tipped back to absorb the rays of the evening sun. Her breath caught as her gaze snagged on the strong line of his throat, the strength of his jaw—Adonis could eat his heart out.

But enough voyeurism…

He turned as she approached and smiled, and for a moment the clock turned back, transported her to two years before, when that smile had quite literally bewitched her, causing her to forget common sense and every promise she'd made herself.

Not this time. This time she had her sensible head on.

So she forced her toes to uncurl and sat down next to him, stretched her legs out and exhaled. 'This is a fabulous place.' She swiped a sideways glance at him. 'And you've surprised me.' *Again.*

'Why?'

'It's not what I expected.'

'What *did* you expect?'

'Something busier—a five-star hotel on the beach, with a nightclub.'

'Is that what you wanted?'

'No.'

'I told you, Sunita, I've changed. Plus, this time needs to be for you and me. No distractions. You wanted to get to know me better. Here I am.'

So he was—and the thought had her reaching for the lime drink she'd ordered.

She needed to focus on the practical—on need-to-know, real-life information.

'I need to know what our marriage would mean on a day-to-day level for Amil. What it will be like for him to grow up in a palace, as a Lycander prince. Right now it feels surreal.'

'The state apartments are a bit more opulent than your average home, I suppose, but otherwise his childhood will be what we make it.'

'Will he go to a nursery?'

'I don't see why not—there will be a certain level of security arrangements, but I can't see a problem with that.'

'And he'll have friends round to play?'

How she'd craved friendship as a child—but there had been no one. Her mixed race heritage, the fact that she was illegitimate, the fact that her mother was a model, had all combined to make school a miserable place of isolation for her. She knew exactly what a solitary childhood could be like, and she didn't want that for her son.

'Yup. Again, subject to security vetting.'

'Is that how it worked for you?'

She sensed the tension in his body.

'It isn't relevant how it was for me,' he said.

He had to be kidding. 'Of course it is. You are a prince who grew up in a palace. You want Amil to do the same. So, did you make friends, have kids round to play? Were you treated differently?'

Discomfort showed as he shifted on his seat, picked his beer up and put it down again without even taking a sip. 'My life...my younger brothers' lives...weren't as straightforward as I hope Amil's will be. There weren't that many opportunities for us to make normal friends. It was better for Axel, because my father sent him to boarding school, and—'

Whoa! 'That is *not* happening to Amil. I will not send him away.'

'I won't rule that out.'

'Yes, you will. I don't care if every Crown

Prince since the Conquest was sent to boarding school. Amil isn't going.'

'That is not why I would do it.' Frustration seeped into his tone. 'In fact, I didn't say I *would* do it. It is simply a possibility I will consider in the future.'

'*No.*'

His voice tightened. 'Different children thrive in different conditions. Axel was educated at boarding school and it didn't do him any harm. I spent a term there and I loved it.'

'In which case, why did you leave?'

'Because my father changed his mind.'

'He must have had a reason.'

'I'm sure he did.'

Despite the even tone of his voice she could sense evasion.

'Do you know what it was?'

'My father's attitude to my education was a little hit and miss. Axel went to boarding school, but the rest of us… We had tutors some of the time, attended a term of local school here and there, or we ran wild. For my father, education wasn't a priority—in the palace or in the principality as a whole. I will change that, but it will take time—that's why I won't rule out boarding school if it is right for Amil.'

'That is *my* decision.'

'Amil is *our* son. *We* will make decisions about his future. Not you or me. Us—together.'

'And what happens if we don't agree?'

'Then we find a compromise.'

'There is no compromise between boarding school and not boarding school. It's black or white. What happens then?'

'I don't know. But we'll work it out.'

'Those are just words. Neither of us has any idea of how to work things out.'

Which was exactly why this was a terrible idea. Co-parenting sucked.

'Fine. Then let's work it out now,' he said.

'How?'

'You tell me exactly why you are so adamant that boarding school is not an option. The truth. My brother loved his boarding school, and the few months I spent there were some of the happiest times of my life. I will not rule it out without reason.'

'I...' Explanations sucked as well, but she could see that she didn't sound rational. 'I'm scared for him. School was an unmitigated disaster for me—because I didn't fit from day one. I was the only mixed race child in my school, and my mother's status didn't help. Plus, quite often she would pull me out of school to go on shoots with her—she had no one to leave me with, you see. I guess I was an obvious target.'

'Were you bullied?'

Although his voice was gentle she could hear an underlying anger, saw the clench of his jaw.

'No. It was much worse. I was ignored. Some girl decided that the best way to treat someone as low down the pecking order as me would be to pretend I was invisible.'

She could still hear it now. The high-pitched voice, so stuck-up and snobbish, the other girls gathering round to listen. 'It is demeaning to even *acknowledge* a dirty girl like her. So we will ignore her. Are we all agreed?'

'My whole experience of school was miserable. The only saving grace was the fact that it wasn't boarding school—that I could go home to my mother. Amil will be different too. He will be royalty—there will be people who are envious of him. I don't want him to be far away and miserable.'

Though in truth there was even more to it than that. There was her bone-deep knowledge that time was infinitely precious—she had had so few years with her mother, but at least they had had the maximum possible time together.

'I don't want him to be far away. Full stop. He is *my* child—I want to see him grow, and I want to be there for him.'

Frederick's hazel eyes studied her expression

with an intensity that made her feel he could read her soul.

Then he nodded. 'OK. You get the casting vote on the boarding school question.'

'Why?' Wariness narrowed her eyes at his capitulation.

'Does it matter?'

'Yes. I need to know that you mean it. That these aren't just words to sweeten the marriage offer.'

'Because you still don't trust me?'

She wanted to—she really did—but how could she when there was so much at stake?

'Let's say it would help if I knew what had changed your mind.'

'You've made me realise why I enjoyed boarding school so much. Why Axel thrived there. It was the opposite to your situation. For us it was an escape from our home life—boarding school was a haven of certainty after the chaos of life at the palace. Somewhere I knew what was what, where I had an opportunity to actually get an education. Our home life was erratic, at best. It won't be like that for Amil.'

Sunita's heart ached at the thought of all those young princes, buffeted by the fallout from their father's chaotic lifestyle. 'No, it won't.'

'And by the time he goes to school I *will* have turned education around in Lycander. Teachers will be better paid, the curriculum will be over-hauled in a good way, and there will be more money injected into schools everywhere.'

As if embarrassed by his own enthusiasm, he leant back with a rueful smile that flipped her heart again. A sure case of topsy-turvy heart syndrome. And it was messing with her head, making the idea of marriage more palatable. *Ridiculous*. Marriage equalled tying herself down, committing herself to a shared life, to a fairy tale ending. The idea hurt her teeth, sent her whole being into revolt.

Only that wasn't true, was it? Horror surfaced at the identification of a tiny glimmer of sparkle inside her that desperately *wanted* a fairy tale ending... Frederick, Sunita and Amil, living happily ever after in a palace. Princess Sunita.

'Penny for your thoughts?' His voice interrupted her reverie.

'They aren't worth it.'

They weren't worth even a fraction of a penny—she had lost the plot and it was time to get it back. This marriage deal wasn't off the table, but there wouldn't be any glimmer of fairy sparkle sprinkled on it.

She looked up as Deepali approached from across the courtyard. 'Your meal is ready. The chef has prepared a selection of traditional Goan food—I trust you will enjoy it.'

Sunita managed a smile even as her brain scrambled around in panic, chasing down that stupid, sparkly bit of her that advocated the ringing out of wedding bells. How had this happened? In a little over twenty-four hours he had somehow persuaded her that marriage was not only a possibility but a sparkly one.

Enough. She had to halt this before this fairy tale place wove some sort of magic spell around her—before that stupid sparkly bit inside her grew.

Frederick studied Sunita's expression as she looked round the dining room. Her eyes skittered over the colourful prints on the white walls, along the simple wooden table, and he could almost hear her brain whirring.

Deepali entered and put their plates in front of them. 'Prawn rissoles,' she said, and Sunita inhaled appreciatively.

'They smell marvellous—and I'm sure they'll taste just as good.'

The middle-aged woman smiled. 'I'll pass on your kind comments to the chef.'

Once she'd gone, Frederick watched as Sunita studied the rissole with more attention than any food warranted, however appetising.

'This looks great.' She popped a forkful into her mouth and closed her eyes. 'Fabulous!

The reason why melt-in-the-mouth is a cliché. Cumin, with perhaps a hint of coriander, and…'

But even as she spoke he knew that her thoughts were elsewhere. There was an almost manic quality to her culinary listing, and he interrupted without compunction.

'So,' he said, 'you avoided my earlier question about what you were thinking.'

Her brown eyes watched him with almost a hint of defiance. 'I was thinking how surreal this situation is—the idea that two people who don't know each other at all could contemplate marriage. It's…mad.'

'That's why we're here—to get to know each other.'

'We can't pack that into two days—most people take years.'

'And there is still a fifty per cent divorce rate.'

'In which case we are *definitely* doomed.'

'Not at all. All those people who take years… they try to fall in love, decide they've fallen in love, expect love to last. Every action is dictated by love. They heap pressure on the whole institution of marriage *and* on themselves. Our approach is based on common sense and on us both getting a deal we think is fair. Two days is more than enough time.'

He leant over and poured wine into her glass. 'In days gone by it would have been the norm.

Throughout Lycander history, rulers made *alliances*—not love matches.'

'Does posterity say whether they worked?'

'Some were more successful than others, but every marriage lasted.'

Until Alphonse had arrived and turned statistics and traditions on their heads.

'For better or worse?' Sunita sounded sceptical.

'I see no reason why we couldn't be one of the better ones—we'd go in without any ridiculous, unrealistic expectations, with an understanding of what each other is looking for.'

'I don't even know what your favourite colour is.'

'Does it matter?'

'I feel it's the sort of thing one should know before they marry someone.'

'OK. Blue.' He raised his eyebrows. '*Now* will you marry me?'

This pulled a reluctant smile from her, but it came with an attendant shake of her head. 'What sort of blue? Royal blue, because it's on the Lycander flag?'

'Nope. Aquamarine blue.'

'Because…?'

'Does there have to be a reason?'

Sunita tipped her head to one side. 'There usually is.'

'So what's *your* favourite colour?'

'Red.'

'Because…?'

'Because it was my mother's favourite colour—I like to think it was her way of sticking two fingers up at the world that had branded her a scarlet woman. She always wore something red—her sari would maybe have a red weave, or she'd wear a red flower, or paint her toenails red. And as for her lipstick collection…'

'You must miss her.'

'I do. A lot.' She looked down at her plate and scooped up the last of her rissole. 'Anyway, why aquamarine blue?'

Reluctance laced his vocal cords—along with a sense of injustice that a question that had seemed so simple on the surface had suddenly become more complex. *Get a grip.* If this was a hoop Sunita had constructed as a prelude to marriage then he'd jump through it—he'd do the damn hula if necessary.

'It's the colour of the Lycander Sea. When life in the palace became too much I'd escape to the beach, watch the sea. It put things into perspective. Sometimes it was so still, so calm, so serene it gave me peace. Occasionally it would be turbulent, and then I guess I'd identify with it. As a child I was pretty sure Neptune lived off the coast of Lycander…'

OK, Frederick, that's enough. More than he'd intended in fact. But there was something about the way Sunita listened—*really* listened—that seemed to have affected him.

She watched him now, lips slightly parted, tawny eyes serious, but as if sensing his discomfort she leant back before she spoke.

'OK, next question. Star sign?'

'Leo.'

'Me too.'

'Is that good or bad?'

'I really don't know. We'd need to ask Nanni—she is an avid believer in horoscopes. Though I'm not sure why. I think her parents had her and my grandfather's horoscopes read to see if they'd be a good match, and the astrologer was confident they were compatible.'

'Were they?'

'I don't think they can have been. From what my mother told me my grandfather was a tyrant and a control freak, whereas Nanni is a kind, gentle woman. But Nanni herself never speaks of her marriage—and never criticises my grandfather. And she still believes in horoscopes.'

'What about you? Do *you* believe in horoscopes?'

'I think there may be something in it, but not enough that you can base your life decisions on them—that's the easy way out, isn't it? You can just shrug your shoulders and blame fate if it all goes wrong. It doesn't work like that—life is about choice.'

'Yes…' Bleakness settled on him—his choices had cost Axel his life. 'But life is also about the consequences of those choices. Consequences you have to live with.'

'Yes, you do. But in this case Amil's future is in *our* hands—he will have to live with the destiny *we* choose for him. And that is hard. But it's not only about Amil. It's about us as well. You and me. That's why this marriage can't work.'

Her chin jutted out at an angle of determination.

Frederick frowned—but before he could respond the door opened and Deepali re-entered the room, followed by a young man pushing a trolley.

'Fish *recheado*,' the young man announced. 'Made with pomfret.'

Deepali's face shone with pride. 'This is my son, Ashok—he is the chef here,' she explained.

'I thought you might want to know about the dish,' Ashok said.

'I'd love to.'

Sunita smiled her trademark smile and Frederick saw Ashok's appreciation.

'The pomfret is stuffed with a special paste. I used chillies, cloves, cumin and lemon. It is a Goan dish, but *recheado* means stuffed in Portuguese.' Ashok smiled. 'And there is also Goan bread, freshly baked. Enjoy.'

Frederick waited until the mother and son had left the room and then he looked at Sunita.

'Why not?' he repeated.

CHAPTER NINE

'WHY WON'T THIS marriage work?'

Frederick's voice was even, his question posed as if the topic under discussion was as simple as a grocery list rather than the rest of their lives.

Sunita took a deep breath and marshalled the thoughts she'd herded into a cogent argument throughout the starter. 'Would you have even considered marriage to me if it wasn't for Amil?'

There was no hesitation as he tipped his hand in the air, palm up. 'No.'

To her surprise, irrational hurt touched her that he didn't have to give it even a second's thought. 'Exactly.'

'But you *can't* take Amil out of the equation. If it weren't for Amil you wouldn't consider marriage to me either.'

'I get that. But it's different for you. I don't *need* to marry anyone. You do, and you need it to be the right person—for Lycander's sake.

A woman like Lady Kaitlin Derwent. I am the *antithesis* of Kaitlin.'

For an insane moment the knowledge hurt. But she was no longer a child, desperately trying to measure up to her half-sisters and always failing. High academic grades, musical ability, natural intelligence… You name it, Sunita lacked it. But in this case she needed to emphasise her failings with pride.

'I haven't got an aristocratic bone in my body, and I don't have the *gravitas* that you need to offer the Lycander people.'

'You are the mother of my son.'

'Your *illegitimate* son. Plus, I was a model. Your father married or was associated with a succession of models, actresses and show-biz people, and all his relationships ended in scandal. Your people will tar me with the same brush.'

'Then so be it. I agree that you do not have the background I was looking for in my bride, but I believe you will win the people over. In time.'

'I don't think I will.' She inhaled deeply. 'For a start, I want to resume my modelling career—and I can't see that going down a storm with the people.'

Or with him. He masked his reaction, but not fast enough—he hadn't taken that into the equation.

'You don't like the idea either?'

'I neither like nor dislike it. I agree it might be problematic for the people to accept, but it's a problem we can work around.'

'But it doesn't *have* to be a problem. Don't marry me—marry someone like Kaitlin… someone with the qualities to be a true consort.'

Even as she said the words a strange pang of what she reluctantly identified as jealousy shot through her veins. *Jealousy? Really?* She didn't even know who she was jealous of. It meant nothing to her if Frederick married someone else. *Nothing.* As for being jealous of Kaitlin—that was absurd.

Sunita forged on. 'You know I'm right. Tell me about your agreement with Kaitlin. What else did she bring to the table apart from her background?'

'This is not a constructive conversation.'

'I disagree. This isn't only about Amil. This is about us as well. Your life and mine. You want to make me a princess—I deserve to know what that entails, what your expectations are. You said it yourself.'

'What I expected from Kaitlin and what I would expect from you are different.'

Ouch. 'In what way?' Ice dripped from her

tone as she forked up a piece of succulent fish with unnecessary violence.

'You are two different individuals—of course I would have different expectations.' Frustration tinged his voice, along with what looked like a growing knowledge that he'd entered stormy waters and was in imminent danger of capsizing.

'Well, I'd like to know what you expected from Kaitlin.' *From your ideal candidate*, her treacherous heart cried out.

'Fine. Kaitlin was brought up for this role— she has dozens of connections, she speaks four European languages, she has diplomacy down pat. I planned to use her as a royal ambassador—she would have played a very public role. I also hoped she would be influential behind the scenes—play a part in turning Lycander round, in shaping policy.'

For Pete's sake! Sunita didn't think she could bear to hear any more. Lady Kaitlin had obviously been on a fast track to royal sainthood, and the role of Lycander princess would have fitted her like a silken glove. Whereas Sunita was more fitted for the lost sock that languished behind the radiator.

The realisation hollowed her tummy and she shook her head in repudiation. 'There you have

it. I think you owe it to Lycander to marry someone else.'

Surely she'd made her case? She understood
that Frederick wanted to be part of Amil's life,
but he *had* to see that Sunita was quite simply
not princess material.

'No.' His voice was flat. 'I have already considered everything you've said. And, inciden-
tally, you and my chief advisor are in complete
agreement. But you are Amil's mother, and that
trumps all other considerations. He is my son.
I want him to live with me—I want him to be
Lycander's Crown Prince after me. I also want
him to live with his mother. So marriage is the
only option.'

'No, it isn't. What if I decide not to marry
you?' He couldn't actually *force* her to the altar.
'You would still be an important part of Amil's
life.'

'Stop!'

'What?' Her stomach plummeted as she saw
the expression on his face—weariness, distaste,
sadness.

'Don't do this.'

'Why not?'

'Because if you don't marry me I will fight
for joint custody.'

Joint custody. The words sucker-punched her.

'You promised that you wouldn't take him from me. You said he needs me.'

'I also told you I will be a real part of his life. What would you suggest? A weekend here and there? He is my son as well.'

'Yes. But you'll marry someone else—have another family.'

'And you think that should make me want Amil less—is that the message you want to give our son?'

'No!'

Damn it—she couldn't think. Panic had her in its grip, squeezing out any coherent thought. All she could think of now was losing Amil for half of his childhood. Of Amil in Lycander with a stepmother—whichever new multilingual paragon of virtue Frederick eventually married—and half-siblings.

History on repeat with a vengeance.

Memories of her own humiliations, inflicted by the hands of her stepmother and her half-sisters—the put-downs, the differentiation, the horror—were chiselled on her very soul. No way would she risk that for Amil.

'I won't agree to joint custody. I *can't*.'

But she could see his point. She had already deprived him of fourteen months of Amil's life—how could she expect him to settle for the occasional week? Regular phone calls and

Skype? Would *she* settle for that? Never in a million years.

She inclined her head. 'All right. You win. I'll marry you.'

It looked as if Princess Sunita was about to enter the land of fairy tales. It was a good thing she knew that happy-ever-afters didn't exist in real life.

CHAPTER TEN

'ALL RIGHT. YOU WIN. I'll marry you.'

The words seemed to haunt his dreams, and by the time the distinctive fluting whistle of a golden oriole penetrated his uneasy repose it was a relief to wake up, hop out of the slatted wooden bed and head for the shower. He could only hope the stream of water would wake him up to common sense.

He had won, and there was nothing wrong with winning—it meant he would have a life with his son, would be able to give Amil his principality. That was *good* news, right?

The problem was Sunita's words had not been the only ones to permeate his sleeping mind. His father's voice had also made a showing.

'Every woman has a price. Find her weakness, exploit it and then you win, Freddy, m'boy.'

He switched off the shower in a savage movement. Time to man up. Yes, he'd won—and that was OK. It was a cause to celebrate—*not* the equivalent of what his father had done. *He* was striving to keep Amil with Sunita full-time. He hadn't destroyed a family—he'd created one. Ergo, he was not his father. It wasn't as if he had *threatened* her with joint custody. It had been the only other option—an option he'd known she would knock back.

Rationally, the facts were undeniable. Sometimes in life you had to choose between the rock and the hard place, and he'd done his best to make the rock a comfortable choice for her. He'd offered her the chance to be a princess—most women would have grabbed the baton and run with it.

End of.

Now it was time to figure out the next step.

He pulled on chinos and a navy T-shirt and headed into the courtyard and the early-morning sunshine.

'Over here.'

He heard Sunita's voice and spotted her sitting under the shade of a tree, simply dressed in a rainbow-striped sundress, sunglasses perched atop her raven hair. Sunlight filtered through the green leaves of the banyan tree, dappling her arms and the wood of the table, lighting up the tentative smile she offered as he approached.

It was a smile that seemed to bathe his skin in

the warmth of relief, pushing away any lingering doubts about his actions.

'Hey.'

'Hey...' He sat down opposite and surveyed the array of fruit. 'Wow.'

'I know, right? It's hard to know where to begin!'

'I'm not even sure I can name them all.'

'*Chiku*, papaya, guava, pineapple, *rambutan*. They all taste different and they are all delicious.'

He reached for a *chiku*—a fruit he'd never heard of. 'It looks like a potato.'

'Wait until you taste it.'

He halved the fruit to reveal pinkish flesh seeded with a mere three black seeds. He scooped out a spoonful and blinked at the intense sweetness.

'Better than cotton candy.'

She smiled, and once again relief touched him.

'About last night...' he said. 'I know marriage isn't your ideal option, but I am very glad you said yes.'

'It isn't, but it *is* the best option on the table and I've decided to make the best of it. Perhaps if I'd been more upfront two years ago we wouldn't be in this mess. But we are, and I'll do my best to be positive about the marriage idea.'

'Our marriage doesn't have to be a mess. I

think we can make this work. For Amil *and* for us.'

A pause, and then she nodded. 'I'll try. So, what's the next step in Project Marriage?'

There was no room for further doubts or any more discussion with his conscience. Project Marriage was what he wanted and what he believed to be right for them all. Yet for some reason he felt restless, as if the beauty of the surroundings was somehow tainted. This was the sort of place where *real* couples should sit and plan their future—couples foolish enough to believe in the concept of love.

'We need a plan, but I suggest we move this discussion to somewhere else. Is there anything you want to see in Goa? We could hit the beach…visit the old quarter…'

In truth he didn't care—he needed to move, to get on with the business of the day away from this tranquil fairy tale setting that seemed to accuse him of having behaved like his father, however much logic told him he hadn't.

Sunita thought for a moment, her tawny eyes dreamy, as if the question needed deeper consideration than it appeared to warrant.

'I'd like to go to the Dudhsagar Falls.'

There was a nuance in her voice he couldn't identify. 'Any reason?'

For a second she hesitated, then she shrugged.

'My parents came to Goa together and they visited the falls. It's one of the few memories my mother ever shared about them both—she said it was important sometimes to remember the happy memories or they would all crumble to dust.'

She picked up a *rambutan*, rolled the lychee-like fruit almost like a dice.

'I'm not entirely sure what she meant, but I'd like to go somewhere she was happy. Even if that happiness was no more than a mirage.'

He had the feeling that right now Sunita missed her mother—and who could blame her? She was about to step into a whole new world that she didn't want to enter.

'I'm sorry you lost her, Sunita.'

'Me too. But I do feel lucky I had her for the time I did.' She hesitated. 'I don't know the details, but I'm guessing you didn't have much time with *your* mum.'

'No.'

Even before the divorce his mother had spent minimal time with him—at least until the divorce proceedings were underway. Then it had all changed, and even now he could remember the glorious happiness his three-year-old self had felt—not the detail, but the joy that finally his mother wanted his company, would hug him, take him out… And then abruptly it had all ceased. She'd gone before the ink had even dried on the papers. The whole 'loving mother' act had been exactly that—an act undertaken to up her settlement.

'I'm sorry.'

'No need. You can't miss what you've never had.'

The words came out rougher than he'd intended, but he didn't want her compassion. He'd got over his mother's abandonment long ago, buried those emotions along with the rest.

Pulling out his phone, he did a check on the falls, scanned the information. 'The falls it is— I'll speak to Security, see how close they can get us. Looks like the official road is closed off because of monsoon season, but I'm sure we can get something sorted.'

'Actually, I wondered if we could do what my parents did and walk along the railway track to get there. Just us—no security. I know they're discreet, but today I'd like to be just Frederick and Sunita—before we get caught up in the reality of being a royal couple.'

The wistfulness in her voice decided him— alongside the fact that, however much he trusted his staff, it made sense to thrash out the details of this marriage in private. Plus if he was being honest with himself, he too wanted to be 'just Frederick and Sunita' for one day. To put aside the burden of ruling and his complex need for this marriage for one day.

'Sounds like a plan.'

Surprise etched her face. 'You're sure?'

'I'm sure. Tell me the route they took and I'll figure it out.'

She grinned. 'I think they came back on a goods train.'

'We can manage that.'

'The Prince and his future consort hopping on a goods train? I like it.'

Her smile broadened and it caught at his heart, causing a sudden unfamiliar tug of hope that perhaps this might all work out.

Sunita glanced up at the sky, and for the first time in the past forty-eight hours her thoughts slowed down as she absorbed the grandeur of the bright grey monsoon clouds.

Most tourists flocked to India in the summer months, but she loved monsoon—always had, even as a child. Loved the drum of the rain, which brought the country much needed water and succour from heat, and lavished verdant green to the trees and fields.

'It doesn't seem possible that there can be so many different shades of green—it makes me wish I could paint, somehow capture all this.' Her outswept arm encapsulated the winding track, the surrounding green and the skies above. 'Photos never seem to catch the reality of it— they look fake, somehow.'

'Then commit it to memory,' Frederick said, putting out a hand to steady her as she stumbled slightly over an awkward rock.

The touch of his hand against hers almost made her gasp out loud, adding an extra level to her already overcrowded senses. In an almost involuntary movement she clasped her fingers around his.

'Like my mother did. She described this walk to me so many times it almost felt like a story.'

Perhaps a real-life fairy tale, in which a moment of happiness had *not* led to a lifetime of happily-ever-after.

'It's odd to think that they walked here once… maybe took the exact same steps we're taking now.' She turned to him. 'You must feel that a lot as a ruler—the idea of history being always around you. Your ancestors' spirits looking over your shoulders.'

For an instant she'd swear a small shiver shot through him, and understanding smote her. Perhaps for him it was the spirit of his older brother that haunted his every move and decision.

Yet his voice was light as he answered, 'I am more worried about current judgement and the opinion of posterity than the line of my progenitors.'

He slowed as they approached a tunnel, half turning for evidence of any oncoming train.

They stepped inside the dark and now it was her turn to shiver at the dank confines. Water trickled down the damp mossy walls and he tightened his grip on her hand. Without thought she moved closer to the strength of his body.

'It's safe. Even if a train does come through there is ample space as long as we keep to the side.'

Yet suddenly it didn't *feel* safe—though it was no longer the train she was worried about. Frederick was too close, and that proximity was playing havoc with her body.

Did it matter any more? They were to be married—their physical attraction could now be acknowledged. The idea jolted a funny little thrill through her—one she short-circuited instantly. Two years ago physical attraction had lambasted her self-control and her pride. No way would she enter *that* thrall again.

As they emerged into sunlight she dropped his hand, under the pretext of tugging her hair into a ponytail, and then turned to him.

'I think we were talking about current judgement and public opinion—and on that topic we need to decide how to announce our engagement.'

For an instant his gaze locked on her hand and then he nodded. 'I think we keep it low-key. I don't want to announce this as a romantic fairy tale—that would be disingenuous, and way too reminiscent of my father's marriages. Every engagement, every wedding was an extravaganza, with proclamations of eternal love.'

'Did he love *any* of them?'

'According to his own criteria he did—but in reality I believe it was little more than lust and an ability to kid himself.'

'Perhaps he did it for children?'

'My father never did anything unless it was for himself.' His tone was factual, rather than bitter. 'But that isn't the point. I don't want to lie and present our marriage as some sort of perfect love story. I'd rather be honest.'

Sunita stared at him. 'That is hardly the most gripping headline—*Prince Proposes to Legitimise Heir.*' Irrational hurt threatened at his reminder that this was the only reason for their union. Well, so be it. 'I don't believe in fairy tales, but I *do* believe in good publicity.'

'So what would *you* suggest?'

'*An old flame is rekindled. Prince Frederick of Lycander and Sunita decide to wed! Both the Prince and his bride profess delight at the prospect of being a real family.*' Her pace increased slightly. 'I mean, that is just off the top of my head—I'm sure your spin people can work on it. We don't have to profess undying love, but anything is better than indifference.'

Admiration glinted in his eyes and warmed her.

'I'd forgotten what a natural you are with publicity. You've definitely not lost your touch.'

'Thank you kindly, good sir. Publicity is an incredibly powerful tool. I agree that we shouldn't lie to your people, but what you are doing is a good, principled action for your son—the people should know that. Of course they'll be interested in a bit of fun and glitz and a celebration too.' She glanced sideways at him. 'Fun is important—for all of us. I want Amil's childhood to be full of fun and joy—I want him to have a happy path through life.'

'So do I.'

'Good. Then let's show your people that. Let's make sure the engagement announcement is honest, but happy. We've decided to do this, so we need to make the best of it.'

With impeccable dramatic timing the skies chose that moment to open up, and before Sunita could do more than let out a warning cry the rain sheeted down in a torrential downpour.

Sunita tipped her face up and let it gush over her, revelling in the sheer force of Nature as it provided one of life's essentials.

Mere moments later the rain ceased. Blue skies replaced the grey, and sudden shafts of bright golden sunshine shot down, illuminating the droplets of water that hung everywhere. The smell of wet earth permeated the air and it seemed impossible not to smile.

'It's as if someone switched the tap off and the lights on,' Frederick said, a note of wonder in his voice as he looked round.

'That would be Varuna, the god of water. Nanni says that he listens to what the frogs say, and when they croak enough he gives us rain.'

'I think I'm going to like Nanni.'

'Of course you are.'

'So I take it your mother's family eventually relented and took her and you back in?'

'No...' Sunita sighed, feeling the familiar ache of regret and sadness. 'I wish that was how it had played out, but it didn't. They didn't relent.'

Even when they knew her mother was dying.

Anger was suddenly added to the mix. Her grandfather hadn't even told Nanni that their daughter was ill—hadn't given her the chance to say goodbye.

'I met Nanni for the first time when I was pregnant with Amil.' She glanced across at him. 'I don't expect your sympathy, but when I found I was pregnant I felt very alone.'

His expression hardened slightly, but to her

surprise she could see an element of frustrated sympathy in his creased brow. 'So you decided to find your mother's relatives?'

'Yes. My mother had left enough information that it wasn't too hard. It turned out my grandfather had died two years before, and Nanni agreed to see me.'

That first meeting was one she would never forget—her grandmother had simply stared at her, tears seeping from her brown eyes, her hands clasped as if in prayer. And then she had stepped forward and hugged Sunita, before standing back and touching her face as if in wonder, no doubt seeing not just her granddaughter but her daughter as well.

'She was overjoyed and so was I. She has never forgiven herself for not standing up to my grandfather, for letting my mother go, and I think she sees me and Amil as her second chance.'

'It isn't always easy to stand up to a partner if he or she has all the power. Your Nanni shouldn't be too hard on herself.'

'I've told her that. My mother didn't blame her either. Nanni was totally dependent on her husband—money, clothes, food, everything—and he made sure she knew it. If she had left with my mother he would have cut her off from the rest of her family, her children...everyone.'

She paused and then turned to him, willing him to understand. 'I won't *ever* let myself get into that position.'

'You won't. Our marriage will be nothing like that.'

'I understand that, but I did mean what I said yesterday—I intend to resume my career. You saw what happened to your mother, your step-mothers. I've seen what happened to Nanni—I will *not* be dependent on you.'

'You won't be. We can set up a pre-nup.'

'In a principality where your word is law? Any pre-nup I sign wouldn't be worth the paper it was written on.'

'OK. You will be paid a salary that goes directly into your personal account—you can move that into another account anywhere in the world.'

'A salary essentially paid by *you*—one you could stop at any moment?'

His lips thinned. 'You really do not trust me at all, do you?'

There was a hint of hurt in his voice, but it was something she could not afford to listen to.

'I can't trust anyone. Think about it, Frederick. What if I decided to take Amil and leave? Would you still pay my salary? What if you turn out to be like your father? What if you fall in love with another woman?' Life had taught her there could never be too many 'what ifs' in the mix. 'Then I'll need money of my own.'

The easy warmth in his hazel eyes vanished, and now his brow was as clouded as a monsoon sky. 'None of those things will happen.'

'That's what you say *now*, but times change— we both know that.'

A shadow flickered across his face and she knew her point had gone home.

'So I must make sure myself that I have enough money in the bank for whatever life throws at me.'

To ensure there was always an escape route— that she would never be trapped like her grandmother had been, as *she* had been as a child.

'That is non-negotiable.'

'Understood.'

'Also, I want to leave Amil with my grandmother when we go back to Lycander.'

'Why?' The syllable was taut. 'Because you think I will snatch him the minute we land on Lycander soil?'

'No. But I won't risk taking him there until we have worked out how our marriage will be received. Also, I can get things ready for him; it will be a big change for him and I'd like to make his transition as easy as possible.'

The idea of not having Amil with her hurt, but she could not—*would* not—risk taking him to Lycander until she was sure of his reception there.

'I'll come back to Lycander with you, and *then* I'll get Amil.'

'OK. But *we* will get Amil.'

She nodded and then there was a silence, broken by a roar in the not so far distance.

'Dhudsagar Falls,' Sunita said. 'We're close.'

By tacit consent they quickened their pace.

CHAPTER ELEVEN

THE SOUND OF the monsoon-inflated waterfalls pounded his eardrums, but even as Frederick anticipated the sight his brain couldn't banish Sunita's expression, the realisation that she still didn't trust him.

Not that he blamed her—after all, his father had used his wealth and power to grind his wives to dust in the courts. All except his mother, who had played Alphonse at his own game and duped him—an act his father had never forgiven her for. Never forgiven Frederick for, come to that. But he wished that Sunita did not think so badly of him. *Enough.* Her opinion shouldn't matter, and in truth she couldn't judge him more harshly than he deserved. But...

His train of thought was broken by her gasp from next to him. 'Any minute now,' she whispered, as they emerged through a tunnel and onto a railway bridge already populated by a few other visitors.

But they had no interest in Sunita and Frederick—because it was impossible to focus on anything other than the waterfalls, both mighty and terrible. No image could do them justice as the four tiers cascaded and roared in torrents of milky-white water, leaping from the edge of towering cliffs and gusting and gushing down the slippery rock slopes.

The spray drenched him but he didn't move, utterly mesmerised by the power and glory of Nature's creation, cloaked in a rising mist that mixed with the shafts of sunlight to create a rainbow of light.

'It's beyond description.'

Frederick nodded and moved by awe, on instinct, he reached out and took her hand in his. He wasn't sure how long they stood there, but it was long enough that the other tourists dispersed, long enough that another group came and went.

And then Sunita shook her head, as if coming out of a trance. 'We'd better go.'

He wondered what she'd been thinking all that time—perhaps she'd imagined her parents standing in the same spot, their thoughts and emotions, their hopes and dreams as they'd gazed at the might of the waterfalls.

They continued their trek along the railway tracks in a silence that he instinctively respected until he motioned to the adjacent forest. 'Shall we explore in there—it looks peaceful?'

'Good idea.' She glanced up at him. 'Sorry I've been lost in thought—it was just such an awe-inspiring sight.'

'It was.' He reached into his backpack. 'Time for food—or is that too prosaic?'

'Nope. I'm starving. And this looks idyllic—if a little damp.'

'I've brought a blanket, and if we spread it here, over this branch, we can perch on it.'

'Perfect.'

She accepted the wrapped sandwiches.

'Goan green chutney,' Frederick informed her. 'I promised Ashok to tell you the exact ingredients. Coriander leaves, coconut, chili and a little sugar and salt.'

Sunita took a bite. 'Glorious. That boy is talented.' She surveyed him. 'So you went to the kitchen yourself? I'm surprised that was allowed.'

'Meaning?'

'Meaning your staff seem to think you shouldn't lift a finger for yourself.'

'I've noticed. I *am* trying to re-educate them—in fact I've given them all the day off today. The problem is my father expected to be waited on hand and foot, and that is what all Lycander staff seem conditioned to do. I even have someone who chooses all my clothes.' He grinned. 'Though, to be fair, Kirsten does a better job of it than I could.'

'Well, for the record, no one is choosing *my* clothes for me. That would drive me nuts. I need to fit my clothes to my mood.'

They ate in companionable silence and then Frederick leant forward, unsure why he felt the need to say his next words, but knowing he had to take heed of the urge to show her that their hopes and dreams didn't have to be built on an altar of falsehood and misunderstanding.

'Sunita?'

'Yes.'

She turned her head and his heart did a funny little jump. Dressed in simple khaki trousers and a red T-shirt, with her hair pulled back in a high ponytail, her features make-up free, she looked absurdly young and touchingly vulnerable.

'I understand why you want to go back to your career, and I understand your need for independence, but let's not go into this marriage expecting the worst. We'll be OK,' he said, even as he realised the ridiculous inadequacy of the words.

'You can't know that.' She lifted her shoulders in a shrug. 'But I appreciate the sentiment.'

'There are some things I *do* know, though. I won't turn into my father.' *Please God!* 'I won't fall in love with someone else and take Amil away from you.'

'You can't know that either.'

'I don't do falling in love, and that will not change. As for Amil, I will not take him from you. You have my word.'

'Words are meaningless.'

Her fierce certainty told him that someone had lied to her with devastating consequence, and increased his need to show her that he would not do the same.

'Sunita, I couldn't do it.' The words rasped from his throat. 'I would have done anything to have a mother. I witnessed first-hand what my father did to his wives, how it affected my brothers. I could not, I *will* not let history repeat itself.'

Her whole body stilled, and then she rose and moved towards him, sat right next to him, so close a tendril of her hair tickled his cheek.

'I'm sorry—I know what it's like to lose a mother through death, but for you it must have been pain of a different type…to know she was out there. And your poor mother…to have lost you like that—I can't imagine how it must have felt. Not just your mother but Stefan's and the twins'.'

For a moment the temptation to let her believe the fiction touched him. To let her believe the false assumption that his mother had been wronged, had spent years in grief and lamenting, that his mother had loved him. After all, he had no wish to be an object of pity or allow the ugly visage of self-pity to show its face.

But as he saw the sympathy, the empathy on her face, he realised he couldn't let her waste that compassion. 'My mother didn't suffer, Sunita.'

'I don't understand.'

'My mother sold me out for generous alimony and a mansion in Beverly Hills. She played my father like a fine fiddle—conned him into believing she would do anything to keep custody of me, would be devastated to lose me. At the time he was still worrying about his popularity—many people hadn't got over the way he'd married my mother mere weeks after his first wife's death. He wanted to hurt her by taking me away—however, he didn't want to come across as the totally cruel husband again, so he offered her a generous settlement and she skipped all the way to the bank.'

'But...' Disbelief lined her face, along with a dark frown. 'How *could* she?'

'With great ease, apparently. Hey, it's OK. I came to terms with it long ago. I didn't tell you because I want to discuss it, or because I want sympathy. I told you because I want you to know that I could never take Amil from you. I know first-hand that a child needs his mother. From my own experience *and* my siblings'. Stefan, Barrett, Emerson—they have all been devastated by the custody battles and having their mothers torn from them. I would not put you or Amil through it. You have my word, and that word is not meaningless.'

'Thank you.' Shifting on the branch so that she faced him, she cupped his face in her hands, her fingers warm against his cheeks. 'Truly. Thank you for sharing that—you didn't have to. And I do believe that right here, right now, you mean what you promise.'

'But you don't believe I'll make good?'

'I…' Her hands dropped to her sides and, leaning forward, she dabbled her fingers in the soil of the forest floor, trickled it through her fingers and then sat up again. 'I don't know.'

She shrugged.

'Perhaps it's my turn to share now. I told you how my father left when he found out my mother was pregnant. He promised her that he loved her, that they had a future.' She gestured back the way they'd come. 'Maybe he said those words at the falls. Hell, maybe they even sat here, in this very forest. But that promise meant nothing. And, you see, that wasn't the only promise he made.'

'I don't understand.'

'He came back.' Her eyes were wide now, looking back into a past that he suspected haunted her. 'When my mother found out she was terminally ill she managed to track him down. She had no one else. And he came, and he agreed to take me in. He explained that he was married, with two other daughters, but he promised—he swore that I would be welcomed, that I would have a family, that he would love and cherish me. He said that he was sorry and that he wanted to make it up to her and to me.'

The pain in her voice caused an ache that banded his chest and he reached out to cover her hand in his own, hoped that somehow it would assuage her hurt.

'So, after she died…' Her voice caught and her fingers tightened around his. 'He came to bring me to England—to my new family.'

Perhaps he should say something, Frederick thought. But he couldn't think of anything—couldn't even begin to contemplate how Sunita must have felt. The loss of her mother, the acquisition of a father she must have had mixed feelings about, the total upending of her life. All he could do was shift closer to her, *show* his comfort.

'It didn't work out. Turned out his promises didn't materialise.'

'What happened?'

'My stepmother and my sisters loathed me—I knew that from the instant I walked into the house.'

A house that must have felt so very alien to her, in a country that must have felt grey and cold and miserable.

'In a nutshell, he pitched me into a Cinderella scenario. They treated me like I was an inferior being.' She made a small exasperated noise. 'It sounds stupid, because it is so difficult to explain, but they made me feel worthless. I ate separately from them, my clothes were bought from charity shops, while my half-sisters' were new, I ended up with loads of extra chores so I could "earn my keep", and there were constant put-downs, constant reminders that I was literally worthless.' Another shrug. 'It all sounds petty, but it made me feel like nothing—worse than invisible. I was visible, but what they could see made them shudder.'

'It doesn't sound petty—it sounds intolerable.' Anger vibrated through him, along with disbelief that people could be so cruel. 'Was your father involved in this?'

'He was more of a bystander than a participant. He was away a lot on business. I did try to explain to him that I was unhappy, that I felt my stepmother didn't like me, but he simply said that I must be imagining it or, worse, he would accuse me of base ingratitude. Which made me feel guilty and even more alone.'

No wonder Sunita found it hard to take people

at their word. Her own father, who had promised to care for her, had instead treated her like muck and allowed others to do the same.

'I'm sorry. I wish I could turn back time and intervene.'

'You can't change the past. And even if you could perhaps the outcome would be worse. Because in the past I got out, I escaped, and I've come to terms with what happened. I can even understand a little why my stepmother acted as she did. She was landed with a strange girl—the daughter of a woman her husband had been unfaithful with, the woman who probably was the love of his life. The gossip and speculation in the community must have been beyond humiliating for her and my half-sisters. So they turned all that anger and humiliation on to me.'

'That doesn't excuse their behaviour, or explain your father's.'

'I think my father was weak and he felt guilty. Guilty over the way he'd treated my mother… guilty that he had betrayed his wife in the first place. And that guilt translated into doing anything for a quiet life. That worked in my favour later on. I got scouted by a model agency when I was sixteen and my father agreed to let me leave home—my stepmother was happy to see me go, sure I'd join the ranks of failed wannabes, so she agreed. I never looked back and I never went back. I never saw them again. The second I could, I sent my father a cheque to cover any costs he might have incurred over the years. As far as I am concerned we are quits. I don't even know where he is.'

So much made sense to Frederick now—her lack of trust, her fears over Amil, her need to be in control. Admiration burned within him that she had achieved so much, was such an amazing parent herself.

These were all the things he wanted to say, but didn't quite know how. So instead he did what he had promised himself he wouldn't do and he kissed her—right there in the middle of the rainforest, with the smell of the monsoon in the air, and the pounding of the waterfall in the distance. He kissed her as if his life and soul depended on it.

Her resistance was brief—a nanosecond of surprise—and then, as if she too were tired of words, of this walk down a memory lane that was lined with sadness, her resistance melted away and her lips parted beneath his.

He tasted the sweet chili tang left by the sandwiches and heard her soft moan. Their surroundings receded. The call of a hornbill, the rustle of the monkeys in the trees above all melted away and left only them, encased in a net of yearning and need and desire.

He pulled her closer, oblivious of the rustle of the blanket, the unwieldiness of the branch they sat on. Nothing mattered but *this*—losing themselves in this moment of sheer bliss as he deepened the kiss, as her hands slipped under his T-shirt so her fingers covered the accelerated beat of his heart.

Who knew what would have happened if a monkey in the tree above hadn't decided to take advantage and scamper down in an audacious bid for the rucksack. It's insistent chatter and the swipe of an overhanging branch brought Frederick back to reality.

A shout from him, a darting movement from Sunita, and the monkey jumped to safety and jabbered at them in indignation.

They met each other's eyes, hers still clouded with desire, and he managed a smile. 'Well saved.'

Then there were no words. They both simply stood there, and he reached out and took her hands in his.

'What now?' he asked.

'I don't know.' She shook her head. 'Yes, I do. Let's walk. And eat and talk. But let's not talk about unhappy things.'

'That sounds good. Only happy topics—all the way back.' He held her gaze. 'And what happens then?'

She stepped forward, stood on tiptoe, and dropped the lightest of kisses on his lips. 'I don't know,' she whispered. 'I really don't.'

For a moment neither did he. Oh, he knew what he *should* do—he should lock this down now. This physical attraction was too intense, too emotional, and he didn't want intensity or emotion to enter their relationship. This marriage was an alliance and he wanted it to last. Succumbing to physical allure, allowing it too much importance, would jeopardise that.

But today he was just Frederick, not the Playboy Prince or the ruler of a principality who had vowed to fulfil his brother's vision. Today they were Frederick and Sunita.

And so he stepped forward and smiled—a smile that was shamelessly predatory and full of promise. 'Then it's lucky that I know *exactly* what to do.'

'What…?' Her voice was even softer than before, her brown eyes wide.

'We're going to walk to the nearest station and catch a goods train, and then take a taxi back to the villa. Then we're going to resume where we've just left off and this time we are not going to stop.' He paused. 'How does that sound?'

He realised he was holding his breath as she tipped her head to one side, and then she smiled a smile that lit her face and ignited a warmth that spread across his chest.

'That sounds perfect. I just hope the goods train is fast.'

Frederick shook his head. 'Anticipation is half the fun.' He held out his hand. 'Let's go.'

'Anticipation is half the fun.'

Sunita wasn't so sure of that. As they walked alongside the train tracks anticipation streamed through her veins, causing her tummy to cartwheel and her pulse-rate to soar. Was *that* fun?

It was hard to tell—her whole body felt tight with need, a yearning that it would now be impossible to quell, and truth to tell she didn't want to. She glanced down at their clasped hands, at the strength of his profile, the jut of his jaw, the lithe assurance he walked with, the whole time aware of his own scrutiny, the desire that warmed his hazel eyes when they rested on her.

They talked—of course they did. Of films and books and politics…of cabbages and kings…but the words seemed to be filtered through a haze of awareness that glistened in the air alongside the sunlit drops of rain that sparkled from the lush leaves.

Their ascent onto a goods train seemed almost surreal as they travelled amidst the bulky cargo, and she gazed out over the variegated greens of paddy fields, the swoop of the Goan valleys, the shimmering grey of the sky, where clouds swelled and perfumed the air with the promise of rain. All the while, even as her senses stored away Nature's munificent beauty, they also revelled in Frederick's proximity, in the knowledge that soon—soon—they would be together, that for a time at least he would be hers.

Careful!

She must not let this get out of perspective, make it into any more than it was. This was a benefit of their marriage deal—a benefit that could be taken or left at will. This was physical—no more no less—and the only reason she was so on edge was because she hadn't felt like this for two years. Not since that night when all her principles had been abandoned and she'd tumbled into bed with him.

'Hey.'

She turned to see his hazel eyes rest on her face.

'You OK? We don't have to do this, you know? We have a lifetime ahead of us…'

But not like this—not as Frederick and Sunita. Today meant something different—she couldn't explain how she knew it, but she knew with soul-wrenching certainty that this was the case.

'I know, but I want this now, today…' She

grinned suddenly. 'There is only so much anticipation a girl can take.'

His answering grin removed all doubts; it was a smile she remembered from two years ago, boyish, happy, and she hadn't seen it once in the past few days.

'It'll be worth it. I promise.'

And as the train slowed to a stop she had no doubt that this was a promise she could rely on.

Twenty minutes later they arrived back at the villa and alighted from the taxi. She glanced around almost furtively, not wanting to meet Deepali or Ashok or anyone. Hand in hand, they practically tiptoed through the garden... And if Deepali did spot them she remained discreetly hidden and they reached Sunita's bedroom safely.

Once inside, she moved to the window and pulled the blackout shutters closed, then turned and moved towards Frederick with an urgency more than matched by his own as he strode forward and pulled her so she was flush against the hard promise of his body.

'The anticipation was great,' she murmured. 'But now it's got old.'

His laugh held a breathless quality. 'Tell me about it!'

And with that he tumbled her back onto the bed, and after that all coherent thought evaporated as sensation took over. The feel of his lips on hers, his taste, his touch against her sensitised skin—all caused her to moan in unabashed joy. His skin under her fingers, his shudder of pleasure, his voice whispering her name, the shucking of clothes, the urgency and the exquisite gentleness, the awe and the laughter and desire such as she had never known, all created a waving, pulsing sense that carried them higher and higher...

Hours later she opened her eyes, realising that she was being gently shaken awake, a hand on her shoulder. *His* hand. She blinked sleepily, and then sat up as the glorious dream dissipated. Frederick stood back from the bed—fully clothed, she noted with a fuzzy disappointment.

'Hey...' she said.

'Hey. Everyone's back—we need to show our faces before they wonder where we are.'

Sunita blinked, tried to compute why it mattered—they were engaged. Surely he wasn't embarrassed. Properly awake now, she propped herself up on one elbow as a sudden awkwardness descended. 'You should have woken me earlier.'

'I thought I'd let you sleep.' Now a small smile quirked into place. 'We expended a lot of energy.'

'So we did.' For a moment relief touched her—maybe she'd imagined the awkwardness.

'But now the day is over and we aren't "just" Sunita and Frederick any more. We are the Prince and his Princess-to-be and we can't repeat this.'

'This?' she echoed, as a spark of anger ignited by hurt flared. 'Define "this".'

His gaze remained steady. 'We can't sleep together again before the wedding. This engagement needs to be seen as completely different from my father's marriages—I don't want the people to believe it is based on physical attraction alone, that their ruler has been influenced by anything other than the good of Lycander.'

'Of course.'

It made perfect sense, she could see that. Of course she could. She could measure every publicity angle with unerring accuracy. This marriage would not play well with Lycander—she herself had pointed that out. So she understood that they needed to downplay their physical attraction and focus on the real reason for their marriage—Amil.

Yet his words felt like a personal rejection, as though beneath his common-sense approach lay reserve, a withdrawal.

The knowledge…the certainty that he regretted the day and its outcome, that he regretted 'this', bolstered her pride, gave her voice a cool assurance. 'I understand.'

After all, he'd made it clear enough. Their physical attraction was a side benefit, a bonus to their marriage alliance, and she would not make the mistake of reading any more into it than that.

'I need to get dressed. Shall I meet you in the gardens?'

For a moment he hesitated, and then nodded and headed for the door. Once he was gone Sunita closed her eyes, annoyed to feel the imminent well of tears. Two years before she'd allowed physical attraction to override common sense, and now it seemed she might have done it again. But no more.

She swung herself out of bed in a brisk movement and headed across the room. Pulling open her wardrobe, she surveyed the contents and settled on a black cold-shoulder crop top over floral silk trousers. A quick shower, a bit of make-up and she was good to go.

Once in the gardens, she spotted Frederick in conversation with Eric, saw the hand-over of a package that Frederick dropped into his pocket before he saw her and walked over.

'Shall we have a quick walk before dinner?' he asked.

'Sure.'

They walked into the sylvan glade, skirting the lily pond, where two brilliant turquoise kingfishers dived, their white 'shirtfront' breasts bright in the dusk.

'I wanted to give you this,' he said, and he reached into his pocket and took out the package, undoing it with deft fingers and handing her the jeweller's box inside. 'We can't announce the engagement without a ring.'

She flipped the lid open and gazed inside. The ring had presence; it glinted up at her, a cold, hard, solid diamond. A discreetly obvious ring that knew its own worth—its multi-faceted edges placed it in the upper echelon of the diamond class. A regal ring—perhaps he hoped it would confer a royal presence on her.

Hell, it was the very Kaitlin of rings.

'Did you choose it?'

For a scant instant discomfort showed, but then it was gone. 'No. Kirsten did.'

The woman who chose his clothes. Of course—who better to choose the correct ring for Lycander's bride?

'I asked her to get it done last night. Is there a problem with it?'

'Of course not.'

In an abrupt movement she pulled the ring out and slid it over her finger, where it sat and looked up at her, each glint one of disdain. The ring wasn't fooled—it knew this was not a worthy hand to rest upon.

Sunita glared down at it as she executed an almost painful mental eye-roll. *Note to self: the ring does not possess a personality. Second note: of course I am worthy.*

She summoned a smile. 'Guess we're all set to go.' Even if she couldn't have felt less ready.

CHAPTER TWELVE

Lycander

THE CAR WOUND up the mountain road. Sunita stifled a gasp and Frederick felt a sudden surge of pride as he saw her reaction to Lycander's castle.

'Holy-moly,' she said. 'It's straight out of a fairy tale. Any minute now Snow White will wave at me from a turret or I'll see Rapunzel climb down a tower.'

Something tugged at his heart as he looked at her—something he couldn't identify and didn't particularly want to. Focus on facts…that was the way to go.

'Believe it or not, this castle has been around for centuries. It started out long ago as a wooden fortress and over the years it has been renovated, added to, and here we have it.'

'It's hard to believe I'm going to live there.'

Equally hard to ascertain her opinion on the fact, he thought.

Sunita subsided into silence as they approached the castle and parked in an impressive paved courtyard, complete with fountains, stone lions and an immense marble sundial.

'I'll give you a proper tour later. For now, if it's all right with you, I'll show you to your rooms—I've asked Giselle Diaz, the housekeeper, to get a set of apartments ready. After the wedding we will move into the state apartments. I'll show you those later—I think you may want to redecorate them.'

Slow down. No need to turn into a tour guide. Come to that, he couldn't help but wonder at the dearth of staff there to greet them. Foreboding touched him—perhaps the no-show was connected to the emergency council meeting he was scheduled to attend right now. Convened to 'discuss'—for that read 'object to'—his marriage.

They reached the apartment suite that would be Sunita and Amil's until the wedding and he scanned it quickly. Clean and polished...welcoming flowers in place. On the surface it all looked fine, but he knew it lacked the extra touches that had abounded the one time Lady Kaitlin had stayed as a Lycander guest. Back then Giselle had been there to greet them, the flowers had been more lavish, the toiletries a tad more luxurious.

Hmm...

'I'll leave you to settle in and I'll be back as soon as I can.'

'Why don't I come with you? My guess is that your council will want to talk to you about our marriage—let's face them together.'

Frederick shook his head. 'I'd rather do it alone. I brokered this marriage—it is my responsibility to explain it to my people.'

A flash of hurt showed in her eyes and then she shrugged. 'As you wish.'

He pushed down the urge to assuage the hurt; this was *his* business and he would deal with it alone.

Fifteen minutes later he looked around the council chamber, which was informally referred to as the tapestry room, due to the needlepoint that lined nearly every centimetre of the walls. The lifework of a princess centuries before, who had toiled whilst her husband had dallied with a string of mistresses.

Each section illustrated a different theme, dominated by war and religion with plenty of fire and brimstone and gore... Presumably it was meant to be an apt backdrop for the discussion of council matters.

'Order!' called one of the council members.

Frederick looked around the table—at Marcus's assessing expression, at the rest of the council's combative stance. 'You requested we meet as soon as I arrived to discuss your concerns. Please enlighten me.'

A middle-aged man rose to his feet. 'This proposed marriage, Your Highness…we do not believe it is a good move.'

'Marcus has kept me apprised of your concerns.' He kept his voice even. 'But this marriage *is* happening.'

'But the people will not like it,' interpolated another council member.

'Sunita is an excellent publicist—I believe she will win them over.'

'How? She is a woman you barely know—a model, the mother of a baby she kept from you— but now that you are on the throne she seems happy to marry you.'

'Shades of your mother…'

'Who said that?' Frederick demanded.

'*I* said it.' The voice came from one of the elder statesmen.

'My relationship with Sunita bears no resemblance to that of my parents.'

'I beg to differ, Your Highness. I was there. Prince Alphonse fell hook, line and sinker for your mother—chased her whilst his wife, the mother of Crown Prince Axel, was dying. Their wedding was an extravaganza pushed forward because the bride was pregnant. Within months of your birth the marriage was floundering; within a few years it had ended in scandal. Your mother played him for a fool.'

White-hot anger roiled inside him. Yet the words were true—a fact he had to face.

'Are you saying that *I* am a fool? What would have happened if you had spoken to my father thus?'

Frederick made a gesture to a guard, who stepped forward without hesitation to a murmur of surprise.

'You would have been marched out and the council would have been shut down until after the wedding.'

He gestured for the guard to stop and rose to his feet.

'But I do not rule as my father did—I have listened to all your concerns and I understand them. Now I tell you this. My marriage to Sunita is to be made in good faith on both our parts. There will be no scandal. There will be no custody battle. This union will endure. This wedding is happening.'

What was he? The Delphi Oracle? But now was no time to exhibit doubt. 'I promise you all that I value your opinion. But you see, ladies and gentlemen, Amil is my son, and if I have a chance to be a father *without* taking my son from his mother then I have to take it. So the wedding will happen and I very much hope to see you all dance at it.'

Further silence, and then Marcus rose to his feet, an enigmatic look on his face. 'I suggest that is the end of this special council meeting.'

As everyone filed out Frederick ran a hand down his face and turned as his chief advisor approached. Frederick shook his head. 'Not now, Marcus. I can't take any more wedding advice.'

The dark-haired man gave a half-smile. 'I wouldn't dare.'

'Now, *that* I don't believe.'

'You should.' Marcus eyed him. 'That is the first time since your ascension to the throne that I have seen you stand up for something *you* believe in.'

'Rubbish. I have stood in this room and fought to convince councillors to support education and tax reform, to close the casinos…'

'I get that. But those were all Axel's policies. This is *your* marriage.'

'Axel would have agreed that I am doing the right thing.'

'Then maybe you and Axel had more in common than I realised.'

For a second his chief advisor's words warmed him—but only for a fleeting second. If Marcus knew the truth he'd never use such words.

Frederick rose to his feet before the urge to

confess overcame common sense and tried to rid himself of the grubby feel of deceit.

'Frederick? I'll support you in this, but you will need to make this work. You need to win the public round.'

'I know.'

Luckily, he knew the perfect person to help with that.

Sunita stared down at the diamond ring that sparkled and glistened and weighted her finger. She looked around the apartment that appeared opulent yet felt oppressive, with its heavy faded gold curtains and the bowls of flowers that, though magnificent, emanated a cloying, gloom-laden scent.

These were showrooms—there should be signs and information leaflets to outline the names of the rich and famous who had stayed within these walls, to document the lives of the painters who had created the looming allegorical creations that adorned them.

The furniture was decorative—but the stripes of the claw-footed chaise longue almost blinded her, and the idea of sitting on it was impossible. As for the bedroom—she'd need a stepladder to get up into a bed that, conversely, seemed to have been made for someone at least a foot smaller than she was.

Well, there was no way she would let Amil live in a showroom, so she needed to make it into a home.

She started to unpack—hung her clothes in the wardrobe, took comfort from the feel of the fabrics, the splash of the colours, every item imbued with memories.

She halted at a knock on the door.

Spinning round, she saw Frederick framed in the doorway, and to her annoyance her heart gave a little pit-pat, a hop, skip and a jump.

'Hi.'

'Sorry, I did knock on the main door.'

'That's fine. How did the meeting go?'

'As well as could be expected. The council understand this marriage. But we need to get the publicity right to prevent a public backlash.'

Sunita moved away from the wardrobe. 'OK. Let's brainstorm.'

Her mind whirred as they moved into the lounge and perched on two ridiculously uncomfortable upholstered chairs.

'We need to make sure the people understand why we have left Amil in India—that it is simply so we can prepare a home for him. I could talk to the local press about my plans to renovate these apartments and the state apartments. I also suggest that before Amil arrives we go on a tour of some of Lycander, so it's clear that I am interested in the country—not just the crown. I won't accept any modelling contracts straight away.'

Even though her agent's phone was already ringing off the proverbial hook.

He rose to his feet, looked down at her with a sudden smile that set her heart off again.

'Let's start now.'

'How?'

'I'll take you on a tour of an olive grove.'

'One of yours?'

For a moment he hesitated, and then he nodded. 'Yes. I'll arrange transport and press coverage.'

'I'll get changed into appropriate clothes for touring an olive grove.' In fact she knew the very dress—a long, floaty, lavender-striped sun dress.

A shot of anticipation thrilled through her.

Stop. This was a publicity stunt—not a romantic jaunt. She had to get a grip. This marriage was an alliance that Frederick had 'brokered'— a word he had used in this very room a mere hour before.

The problem was, however hard she tried— and she'd tried incredibly hard—that anticipation refused to be suppressed by logic or any other device she could come up with.

Perhaps it was simply to do with the glorious weather, the cerulean blue sky, the hazy heat of the late August sun whose rays kissed and dappled the rolling hills and plains of the Lycander countryside. She could only hope it was nothing to do with the man who sat beside her in the back of the chauffeured car.

'So, where exactly are we going?'

'The place where it all started—the first olive grove I owned. It was left to me by a great-uncle when I was twenty-one. I visited on a whim and—*kaboom!*—the whole process fascinated me. The family who lived there were thrilled as my great-uncle had had no interest in the place—they taught me all about the business and that's how Freddy Petrelli's Olive Oil came into being. I expanded, bought up some smaller businesses, consolidated, and now our oil is stocked worldwide.'

'Are you still part of the company?'

'I'm still on the board, but by necessity I have had to delegate.'

'That's pretty impressive—to take one rundown olive grove and turn it into a multi-million-dollar business in a few years.'

'You turned yourself into one of the world's most sought-after supermodels in much the same time-frame. That's pretty impressive too.'

'Thank you—but it didn't feel impressive at the time.' Back then she'd been driven. 'I *needed* to succeed—I would not let my family see me fail. I wanted them to know that they had been wrong about me. I wanted to show them I was my mother's daughter and proud of it.'

At every photo shoot, she'd imagined their faces, tinged a shade of virulent green as they opened a magazine to see Sunita's face.

'That's understandable—and kudos to you for your success. You have my full admiration and, although it may not be politically correct, I hope they choked on envy every time they saw your picture.'

She couldn't help but laugh as a sudden warmth flooded her—it had been a long time since anyone had sounded so protective of her.

Before she could respond further the car came to a halt—and right after that they were mobbed. Or that was what it felt like. Once she had alighted from the car she realised the 'mob' actually consisted of four people—a middle-aged couple, a youth and a young girl—all of whom broke into simultaneous speech.

'The crop has been excellent this year. The olives—they will be the best yet. And last year's olive oil—the gods have blessed it, Freddy!'

'It has been too long, Frederick, too long— how can you have not been here for so long? And why didn't you tell me of this visit earlier? I would have prepared your favourite dishes. Now. Bah… All we have is what I have had time to prepare.'

'Thanks so much for the links to the bikes. Oil, gears, helmets…'

'Frederick, I've missed you! Why haven't you visited?'

There was no mistaking the family's happiness at seeing him, and as Sunita watched Frederick contend with the barrage of comments his smile flashed with a youthful boyishness.

'Pepita, Juan, Max, Flo—I'd like to introduce you to Sunita…my fiancée.'

For a moment the silence felt heavy, and Sunita could feel her tummy twist, and then Pepita stepped forward.

'Welcome, Sunita. It is lovely that Frederick has brought you here. We have all been reading the papers—every article. The little *bambino* looks adorable.'

As she spoke Pepita swept them forward towards a whitewashed villa. Terracotta tiles gleamed in the sunshine and trees shaded the courtyard outside.

'Come—lunch is all ready, Alberto, sort out the drinks. Flo, set the table, Max, come and help me serve.'

'Can I help?' asked Sunita.

'No, no, no. You and Frederick go and sit.'

Within minutes, amidst much debate and chat, food appeared. A bottle of wine was opened, tantalising smells laced the air and Pepita beamed.

'Come and serve yourselves. Frederick, you have lost weight—I want you to eat. They are not your favourite dishes, but they are still good.'

'Pepita, everything you cook is good.'

Sunita shook her head. 'Nope. Everything you cook is *amazing*.'

It truly was. The table was laden with a variety of dishes. Bite-sized skewers that held tangy mozzarella, luscious tomatoes that tasted of sunshine and basil. Deep-fried golden rounds of cheese tortellini. Freshly baked bread with a pesto and vinegar dip that made her tastebuds tingle. Baked asparagus wrapped in prosciutto. And of course bowls of olives with a real depth of zing.

But what was truly amazing was the interplay between Frederick and the family—to see him set aside his role of ruler, to see him morph back to the man he had been before tragedy had intervened and changed his life path.

There was conversation and laughter, the clatter of cutlery, the taste of light red wine, the dapple of sunshine through the leaves causing a dance of sunbeams on the wooden slats of the table.

Until finally everyone was replete and this time Sunita insisted. 'I'll help clear.'

Frederick rose as well, but Pepita waved him down. 'Stay. Drink more wine. I want to talk to your fiancée alone.'

A hint of wariness crossed his face, but clearly he didn't feel equal to the task of intervention. So, plates in hand, Sunita followed Pepita to a whitewashed kitchen, scented by the fresh herbs that grew on the windowsill. Garlic hung from the rafters, alongside copper pots and pans.

'It is good to see Frederick here,' Pepita ventured with a sideways glance. 'And now he is a father.'

'Yes.' Sunita placed the plates down and turned to face the older woman. 'I know you must be angry at what I did, but—'

'It is not my place to be angry—this is a matter for you to sort out with Frederick...a matter between husband and wife. I want to tell you that I am worried about him. Since his brother's death we have barely seen him—all he does now is work. I know that he avoids us. There is a demon that drives him and you need to get rid of it.'

'I... Our marriage isn't going to be like that, Pepita...'

'Bah! You plan to spend your lives together, yes? Then that is your job.'

Her head spun as the enormity of Pepita's words sank in. She *was* planning to spend her life with Frederick—her *life*. She wouldn't have another one; this was it.

Closing her eyes, she forced her thoughts to

centre, to concentrate on the here and now. 'I'll do what I can,' she heard herself say, wanting to soothe the other woman's worry.

Frederick looked up as Sunita emerged from the house. Her face was slightly flushed from the sun, her striped dress the perfect outfit for a sunny day. Her expression looked thoughtful, and he couldn't help but wonder what Pepita had said—though he was wise enough to have no intention of asking. He suspected he might not like the answer.

Guilt twanged at the paucity of time he had given to this family—people he felt closer to than his 'real' family.

He turned to Juan. 'Is it all right if I take Sunita on a tour?'

'Why are you asking me? It is *your* grove, Frederick—I just tend it for you. Go—show your beautiful lady the most beautiful place in the world.'

Sunita grinned up at him as they made their way towards the fields. 'They are a lovely family.'

'That they are.'

'And it's good to see your royal authority in action.'

'Sarcasm will get you nowhere—but you're right. Pepita wouldn't recognise royal author- ity if it rose up and bit her. To her I am still the twenty-one-year-old they taught the olive oil business. Right here.'

Sunita gave a small gasp, her face animated as she gazed ahead to where majestic lines of evergreen trees abounded. Olives clustered at the ends of branches clad with silver leaves that gleamed in the sunlight.

'The colour of those leaves—it's like they're threaded with real silver.'

'That's actually the colour of the underside of the leaf. When it's hot the leaves turn light-side up to reflect the sun. When it's cold they turn grey-green side up to absorb the sun.'

'That's pretty incredible when you think about it.'

'The whole process is incredible. The olives are growing at the moment. They won't be ready to harvest for another few months. You should be here for the harvest. It's incredible. The green table olives get picked in September, October, then the ones we use for oil from mid-November, when they are bursting with oil. It is exhausting work. You basically spread a cloth under the trees to catch the olives and then you hit the trees with sticks. The harvest then gets carted off to the mills—which is equally fascinating. But I won't bore you with it now.'

'You aren't boring me. Keep going. Truly.'

Her face registered genuine interest, and

so as they walked he talked and she listened. They inhaled the tang of the olives mingled with the scent of honeysuckle carried on the gentle breeze, revelled in the warmth of the sun and the lazy drone of bees in the distance.

It was impossible not to feel at peace here. Impossible not to note Sunita's beauty—her dark hair shining with a raven sheen in the sunlight, the classic beauty of her face enhanced by the surroundings—and it took all his willpower not to kiss her. That would be a bad move.

She looked up at him. 'I can see why you fell in love with this place—it has a timeless quality.'

'A few of these trees have been here for centuries.'

'And in that time history has played out... generations of people have walked these fields, beaten the trees with sticks, experienced joy and sadness and the full gamut of emotion in between. It gives you perspective.' She gave him a sideways glance and took a quick inhalation of breath. 'Maybe you should come here more often.'

'Because you think I need perspective?'

'Because Pepita misses you.'

'Did she say that?'

'No, but it's pretty obvious. I'd guess that you miss them too.'

'I don't have time to miss them. In the same

way I don't have time to come down here—my days in the olive oil industry are over, and I've accepted that.'

'That doesn't mean you can't visit more often.' She stopped now and turned to face him, forced him to halt as well. 'No one would grudge you some down-time. And this place *means* something to you.'

That was the trouble—this place took him off his game, distracted him from his mission, reminded him of a time untainted by guilt, of the man he had once been and could never be again. When Axel had been alive.

Yet he had brought Sunita here today—*why?* The reason smacked into him. He'd succumbed to temptation—one more day of 'just Frederick and Sunita'. *Foolish.* 'Just Frederick and Sunita' didn't exist.

'Yes, it does. It represents the past. A part of my life that is over. For good.'

Reality was the crown of Lycander and the path he had set himself. Axel had died—had been denied the chance to rule, to live, to marry, to have children. The only thing Frederick could do now was honour his memory—ensure his vision was accomplished. Ensure the monarchy was safe and Lycander prospered.

'It's time to get back to the palace.'

CHAPTER THIRTEEN

Two weeks later

SUNITA GAZED AROUND the transformed apartments with satisfaction. It hadn't been easy, but the spindly chairs of discomfort, the antique non-toddler-friendly glass tables, the dark gloomy pictures were all gone—and she didn't care if they *were* by museum-worthy artists. Mostly it hadn't been easy because of the intense levels of disapproval exhibited by nearly every single member of staff she'd asked for help.

In truth, Sunita quite simply didn't get it—she hadn't expected instant love or loyalty, but this condescension hidden behind a thin veneer of politeness was both horrible and familiar. It made her feel worthless inside—just as she had in her stepmother's home.

Giselle Diaz, the housekeeper, looked down her aristocratic nose at her, Sven Nordstrom, chief steward, somehow managed to convey utter horror, and the more junior members of staff had taken their cue from their superiors. Whilst they listened to Sunita's instructions, they did so with a frigid politeness that made her quake.

But she'd stuck to her guns, had ransacked the palace for *real* items of furniture, and tucked away in nooks and crannies she'd discovered some true treasures.

Old overstuffed armchairs, ridiculously comfortable sofas…and now she and Amil had a home, a haven.

Sunita gazed at her son. They had brought him back from Mumbai ten days before and he had settled in with a happiness she could only envy. With a smile, he crawled across the floor and she scooped him up onto her lap.

'What do you think, sweetheart?' She showed him two different fabric swatches. 'Do you like this one or this one? For your new nursery when we move to the state apartments.'

'Dabadabad!' Amil said chattily.

'Shall we ask Daddy? That's a good idea, isn't it?'

Frederick would arrive at any moment— every day without fail he was there for Amil's breakfast and tea, and for bedtime. Otherwise he worked.

Ever since the olive grove he'd been distant,

as if he'd built a wall of transparent glass that she couldn't penetrate. He was polite, kind and unfailingly courteous, and it made her want to scream. It also made her wonder what demon drove him to spend nigh on every waking hour in the council room, closeted with advisors, lawyers, education experts or engrossed in legal and constitutional tomes that dated back centuries.

Her reverie was interrupted by the familiar knock on the door.

'Come in.'

Frederick entered and, as happened each and every day, her heart fluttered and she noted the lines of tiredness around his eyes and wished she could smooth them away.

'Adadadadaa!' Amil said, and if she'd blinked she'd have missed the smile that lit Frederick's face—one of pure, unaffected joy—before his expression morphed back to neutral.

'Good evening, Amil. And what have you got for tea today?'

'He has lasagne with carrot sticks. Prepared by his very loving, very lovely potential new nanny.'

Satisfaction pumped a fist inside her as she saw his eyebrows snap together—that had at least got his attention.

'Nanny? You didn't tell me you'd chosen one from the list I gave you.'

The list that had chilled her very bone marrow—a list of extremely qualified, excessively expensive women.

'That's because she isn't on the list. But maybe we could discuss this once Amil is in bed.'

'Sure.'

'Then let's get tea underway.'

She headed to the kitchenette and soon had Amil seated in his high chair.

As she did every day, she asked, 'Would you like to feed him?'

He replied as he did every day. 'No. I'm good, thanks. It looks tricky, and I don't know how you manage to get more food into him than ends up elsewhere.'

True enough, meal times weren't the tidiest of processes—and equally true she *had* worked out a dextrous method of spooning in maximum food—but still… She wasn't sure that his reluctance stemmed from fastidiousness. As for worrying that Amil wouldn't get enough to eat, that didn't ring true either—as she had pointed out, he could always have a second helping.

Perhaps he didn't like the idea of being watched and judged.

'I can go into the lounge whilst you feed him, if you like?'

'I'd prefer it if *you* fed him, if that's OK?'

'Of course.'

Only it wasn't OK. Not really.

Just like it wasn't really OK that Frederick didn't engage in bathtime, didn't take Amil onto his lap for his bedtime story. If it were any other man she would suspect that he didn't care, that he was going through the motions. But that didn't make sense. Frederick had fought tooth and nail to be a full-time father to Amil—risked his throne, defied all advice, was willing to take a less than ideal bride.

'Say goodnight to Daddy.'

The little boy gurgled happily and she walked over so that Frederick could give him a kiss.

'See you in a minute.'

Fifteen minutes later she tiptoed from Amil's room and entered the lounge—then stopped on the threshold and cursed under her breath.

Damn. She'd left her sketchbook open on the table—worse, she'd left it open, so she could hardly blame Frederick for sitting there and studying the page.

'Did you do this?'

'Yes.'

There was little point in denial—it wasn't as if he'd believe that *Amil* had drawn a ballroom dress or an off-the-shoulder top.

'They're good.'

'Thank you—they're just sketches…doodles, really. You know how much I love clothes.'

'These look like more than doodles—you've written notes on fabric and cut. How many of these sketchbooks have you got?'

'It doesn't matter.' No way would she confess the number. 'I've always enjoyed sketching and I've always loved fashion. Ever since my mum took me on a photo shoot with her—I loved the buzz, the vibrancy, but most of all I loved the clothes. The feel, the look, the way they could totally transform a person. Sounds mad, maybe, but I think clothes have power.'

His gaze returned to the sketchbook. 'Have you ever thought about fashion design?'

'No.'

That might be a little bit of a fib, but she didn't really want to discuss it. Her sketches were private—she'd never shown them to anyone and she wasn't about to start now.

'It's just a hobby. I think my forte is wearing clothes, not designing them.'

Moving forward, she removed the sketchpad and closed it with a finality she hoped he would apply to the whole topic.

'Anyway I wanted to talk to you about my nanny idea.'

In truth, she wasn't that keen on a nanny—but she could see that if she planned to model and fulfil her commitments as a Lycander consort then it would be necessary.

'Go ahead.'

'I want to give Gloria Russo the role.'

Frederick frowned. 'I thought she worked in the palace kitchens.'

'She does. That's where I met her. I went down there to sort out how it all works—whether I am supposed to shop, or food is delivered, how and where and when I can cook Amil's food… Anyway, Gloria was really helpful.'

Which had made a novel change from every other staff member.

'She only joined the staff recently, but obviously she has been security vetted.'

'So she used to be a nanny?'

'No.'

'All the people on my list have been trained as a nanny—they have extensive qualifications and experience.'

'So does Gloria—she has four grown-up children. And, most important, Amil loves her already.'

'Amil needs a *proper* nanny.'

Frustrated anger rolled over her in a tidal wave—a culmination of being patronised all week and a need to make her own presence felt in a world she didn't fit into. *Again*.

'*Will* you get your royal head out of your royal behind? Gloria will *be* a proper nanny. She knows how to keep him safe and she knows how to provide love and security and fun. She makes him laugh, but she will also make sure he listens. At least agree to meet her and see her with Amil.'

'As long as *you* agree to meet two people from the list. Then we will make the decision.'

'Deal. You'll like Gloria—I'm sure you will. She is kind and she's down to earth and she's fun. Fun is important.' Something Frederick seemed to have forgotten. 'You must remember that—you used to be the Prince of Fun.'

'That was a long time ago.' His tone implied a lifetime rather than mere months.

'Do you miss it? That lifestyle?'

When there had been a different woman in his bed whenever he wanted, and all he'd had to worry about was where the next bottle of champagne was coming from.

'That life feels like it belonged to someone else. So, no, I don't miss it.'

'I know what you mean. My life before Amil seems surreal sometimes, but there are parts of it that I want to retain—I still love clothes, I'm still Sunita.'

Whereas the Frederick of before—apart from the occasional glimpse—seemed to have vanished completely, remorselessly filtered out by grief and the weight of a crown.

'I know that you have taken on a huge responsibility, and of course you need to take that seriously. But there are aspects of the old Frederick that you should keep. The ability to have fun, to laugh and make others laugh.'

'I've had my quota of fun.' He rose to his feet. 'I have a meeting with Marcus now, so...'

'You have to go.'

Sunita bit her lip, told herself it didn't matter. Why should it? Their marriage was an alliance made for Amil's sake—any desire for his company was both ridiculous and clearly unreciprocated.

'Don't forget about tomorrow. We have a family day out scheduled.'

'It's in the diary.' He looked down at her. 'You are sure you don't want to tell me where we're going?'

'Nope. It's a surprise.'

It was an idea she knew the press would love—the fiancée taking her Prince to a surprise destination with their son. A way of emphasising to the people that their Prince and his Princess-to-be had changed and their party lifestyle was well and truly over.

She smiled at him. 'It will be fun.'

For a moment she thought he would return the smile, but instead he merely nodded. 'Goodnight, Sunita.'

'Goodnight.'

There it was again—that stupid yearning to ask him to stay.

Not happening.

The door clicked behind him as her phone buzzed. Her agent.

'Hi, Harvey.'

'Hey, sweetheart. We need to talk.'

Frederick checked the weather forecast as he approached Sunita's apartments. A sunny and cloud-free day—a typical late-summer day in Lycander, perfect for a 'family day out'. The words had an alien twist to them—family days out had been few and far between in his childhood. And now both anticipation and an irrational fear tightened his gut.

Fear at the level of anticipation, and the knowledge that too much of it was tied up in Sunita, was mixed with the fear of messing it up with Amil. Somehow he had to get these fledgling emotions under control—work out which were acceptable, which he needed to nourish to be a good father and which he needed to stifle before they got out of hand.

He could *do* this—he was a past master at emotional lockdown and he would work it out. He would achieve the balanced, calm marriage alliance he wanted.

Pushing open the door, he entered. Sunita

smiled at him and his breath caught. *Beautiful*—there was no other word to describe her. She was dressed in flared, delicately embroidered jeans and a simple dark blue sleeveless top, sunglasses perched atop her head and her hair tumbling loose in a riot of waves. Her vibrancy lit the room—a room that she'd made home.

Clutter without untidiness gave it a feeling of relaxed warmth, as did the overstuffed armchairs and sofas that she had commandeered from somewhere in the palace to replace the antique showcase furniture.

'You ready?' he asked.

'We're ready—aren't we, Amil? Look, it's Daddy.'

Frederick turned his head to look at Amil, who waved his favourite toy cat at him in greeting. And then he twisted, placed his hands on the sofa cushion and hauled himself up so he was standing. He turned and—almost by mistake—let go, tottered for a moment, found his balance, and then took a step…and another step…and another until he reached Frederick and clutched at his legs for balance.

He looked down at his son—his son who had just taken his first steps. Amil had a look of utter awe on his face, as the life-changing knowledge had dawned on him—he could walk! Frederick's chest contracted with pride and wonderment as Amil turned and tottered back, with each step gaining confidence, until he reached the sofa and looked to Sunita for confirmation of his cleverness.

Sunita let out a laugh of sheer delight and flew across the room, scooped Amil up and spun him round. 'What a clever boy!' she said as she smothered him in kisses, before spinning to a halt right in front of Frederick.

Something twisted in his chest as he looked at them—a strand of emotion almost painful in its intensity. Sunita's face was slightly flushed, her tawny eyes were bright with happiness and pride, and it filled him with yearning. Like a boy locked out of the sweetshop for ever—doomed always to gaze at the sweets he could never, ever taste.

He forced a smile to his lips and hoped it didn't look as corpse-like as it felt. 'He is a very, *very* clever boy.'

Amil beamed at him and that strand tightened.

Frederick cleared his throat, turned slightly away. 'So, what's the plan?' he asked.

'First up we'll do a little press conference.'

'How can I do a press conference if I don't know where I'm going?'

'You'll have to let me do the talking.' She grinned at him. 'Don't look so worried. There are a million royal duties I am *not* equipped for, but I *am* good with the press.'

The words, though casually stated, held a shade of bitterness, but before he could do more than frown she had headed for the door.

Once in the palace grounds, with a knot of reporters, Amil proudly demonstrated his new ability and she seemed totally at ease.

'Hi, all. I've decided the Prince needs a day off—because even a ruler needs some downtime. So we are off on a family day out—I promise I'll take some pics, which I'll pass on to you. As I'm sure you all appreciate we are still a new family, so we'd appreciate some privacy.'

'And what about you, Sunita? Do *you* deserve a day off? Isn't it true you're headed back to the catwalk?'

'That's the plan—but I'll let you know more about that when I know the details.'

'Don't you feel you should focus on your role as Princess, like Lady Kaitlin would have?'

Frederick felt her tense, sensed her palpable effort to relax. 'Lady Kaitlin and I are two different people, so we are bound to approach the role differently.'

He stepped forward. 'Hey, guys, any questions for me? I'm feeling left out.'

The tactic worked and fifteen minutes later he wound the meeting down. 'OK, everyone, fun though this is, we need to head off.'

Sunita delivered the parting shot. 'Amil, wave

to the nice reporters. That would be that one...
that one...and that one.'

Not the one who had brought up Lady Kaitlin.

Laughter greeted this, and Sunita smiled.
'Have a great day—and, as I said, the pics will
be with you soon!'

Once they were alone, Sunita nodded towards
one of the palace cars. 'Hop in. We're off to
Xanos Island.' She paused. 'I hope that's OK?
Marcus suggested it.'

That surprised him. 'I'd like to take Amil
there. Eloise used to take us. Me and Stefan and
Axel, and Marcus as well, because he and Axel
were best friends.'

'Eloise is Stefan's mother, correct?'

'Yes. She came after my mother, and she truly
tried to be a good mother to Axel and me.'

'What happened?'

'Same old story. My father decided to divorce
her and it all disintegrated into an awful custody battle. It ended up that she was allowed occasional visitation with Stefan, and only if she
agreed not to see Axel or me.'

'That must have been tough on you and Axel.'

'Yes.'

The comprehension that he wouldn't see
Eloise again, witnessing Stefan's fury and pain—
it had all hurt. The emotions had been painful,
until he'd locked the futile grief down, figured

out that love could never be worth this type of loss. First his mother, then Eloise—never again.

'But let's not spoil the day—I want to make this a happy memory for Amil.'

'Then let's do that.'

Her smile lit the very air and he forced himself to turn away from it before he did something stupid. Something emotional.

The car slowed down at the small Lycandrian port, and minutes later they boarded the motor boat that would ferry them across. Frederick watched Amil's curiosity and joy at this unprecedented adventure, listened to Sunita as she broke into song and encouraged the Captain to join in.

Closing his eyes, he inhaled the salty tang of the sea breeze, absorbed the sound of her song and the cry of the curlews as they soared in the turquoise blue of the unclouded sky.

Once at the island they alighted and headed over the rocks to the beach, where Sunita produced buckets and spades and a large tartan blanket that she spread out over the sun-bleached sand. Amil sat down and waved a spade with energetic abandon and Sunita grinned as she handed another one to Frederick.

'Right. I thought we'd try and do a sand replica of the Lycander palace—but I think the hard work may be down to you and me.'

The next hours skated by, and Frederick knew

he would add this to his list of happy memories. Preventing Amil from eating sand, building turrets and digging moats, the good-natured bickering over the best way to make the walls secure, Amil walking in the sand, eating the picnic prepared by Gloria—it was all picture-perfect.

'We need to consider Gloria for the role of royal picnic-maker as well as royal nanny.'

'So you really *will* consider her? Give her a fair chance?'

'Of course. But in return I'd like *you* to do something.'

Tawny eyes narrowed. 'What's that?'

'Put a portfolio together and send it to a fashion design college. Or talk directly with an actual fashion house. You said to me that Gloria doesn't need formal qualifications to do the job—maybe you don't either.'

The sketches he'd seen the previous day showed talent—he knew it.

'Those sketches had a certain something about them that I suspect is unique to you—I think you should get them checked out.'

'I'll think about it,' she said, in a voice that was clearly humouring him as she pulled the picnic basket towards her. 'Mmm...chocolate cake.'

Frederick raised his eyebrows. 'Are you trying to change the subject?'

'How did you guess?'

'I'm bright like that. Come on, Sunita, why not send them off? What have you got to lose?'

She looked away from him, out over the dark blue crested waves that sculled gently towards the shore, towards the horizon where a ferry chugged purposefully.

Turning back to him, she shrugged. 'I could lose something precious. Those sketches kept me sane—they were my own private dream growing up. They represented hope that I wasn't totally worthless, not utterly stupid. I don't want to expose them to anyone. I've never shown them to a soul.'

And he understood why—she would have been terrified of the comments from her step-mother or sisters...she would have hoarded her talent and hugged it tight.

'Then maybe now is the time. Don't let them win—all those mean-spirited people who put you down. You've already proved your success to them.'

She shook her head. 'Only through model-ling—that's dumb genetic luck, plus being in the right place at the right time. Fashion design requires a whole lot more than that.'

'I understand that you're scared—and I know it won't be easy—but if fashion design is your dream then you should go for it. Don't let them hold you back from your potential. Don't let what they did affect your life.'

'Why not? *You* are.'

He hadn't seen that one coming. 'Meaning…?'

'I think you're scared too.'

'Of what?'

'Of bonding with Amil.'

The words hit him, causing his breath to catch. She moved across the rug closer to him, contrition written all over her beautiful face.

'Sorry. I didn't mean that in a tit-for-tat way, or as an accusation. It's just I see how you look at him, with such love, but then I see how you hold back from being alone with him. You won't even hold him and I don't understand why.'

There was silence, and Frederick knew he needed to tell her. He couldn't bear her to believe he didn't *want* to hold his son.

'Because of dumb genetic luck.'

'I don't get it.'

'When I was eighteen I went to see my mum. I hoped that there had been some mistake—that my father had lied to me, that she hadn't really abandoned me and that there was some reason that would explain it all away. It turned out there was—she explained that she quite simply lacked the parenting gene.'

'She *said* that?'

'Yes.'

Her hands clenched into fists and her eyes

positively blazed. 'Is *that* what you're scared of? That you lack the parenting gene?'

His gaze went involuntarily to his son, who lay asleep on the blanket, his bottom in the air, his impossibly long lashes sweeping his cheeks.

He couldn't answer—didn't need to. Even he could hear the affirmation in his silence.

'You don't.' Sunita leaned across, brushing his forearm in the lightest of touches. 'I can see how much you love Amil. You are not your mother *or* your father. You are *you*, and you are a great father—please believe that. Trust yourself. I promise I trust you.' She inhaled an audible breath. 'And I'll prove it to you.'

'What do you mean?' Panic began to unfurl as she rose to her feet.

'He's all yours. I'll see you back at the palace much later. Obviously call if there is an emergency—otherwise the boat will return for you in a few hours.'

Before he could react she started walking across the sand. He looked at his sleeping son. Obviously he couldn't leave Amil in order to chase after Sunita, which meant...which meant he was stuck here.

His brain struggled to work as he watched Sunita disappear over the rocks. If he called after her he would wake Amil, and that was a bad plan. He stilled, barely daring to breathe as he watched the rise and fall of his son's chest. Maybe Amil would sleep for the next few hours...

As if he should be so lucky.

With impeccable timing Amil rolled over and opened his eyes just as the sound of the boat chug-chug-chugging away reached his ears. He looked round for his mother, failed to see her, and sat up and gazed at his father. Panicked hazel eyes met panicked hazel eyes and Amil began to wail.

Instinct took over—his need to offer comfort prompted automatic movement and he picked Amil up. He held the warm, sleepy bundle close to his heart and felt something deep inside him start to thaw. The panic was still present, but as Amil snuggled into him, as one chubby hand grabbed a lock of his hair and as the wails started to subside, so did the panic.

To be replaced by a sensation of peace, of unconditional love and an utter determination to keep this precious human being safe from all harm—to be there for him no matter what it took.

CHAPTER FOURTEEN

SUNITA GLANCED AT her watch and then back at the sketchbooks spread out on the table. Thoughts chased each other round her brain—about Frederick and Amil; she hoped with all her heart that right now father and son had started the bonding process that would last a lifetime.

Her gaze landed back on the design sketches and she wondered whether Frederick had been right—that fear of rejection and self-doubt stood in her way. Just as they did in his. Could she pursue a dream? Or was it foolish for a woman with no qualifications to put her head above the parapet and invite censure? Or, worse, ridicule.

Another glance at her watch and she closed the books, piled them up and moved them onto a shelf. Instead she pulled out a folder—design ideas for the state rooms, where they would move after the wedding.

The door opened and she looked up as Frederick came in, Amil in his arms. There was sand in their hair and identical smiles on their faces. *Keep it cool...don't overreact.* But in truth her heart swelled to see them, both looking so proud and happy and downright cute.

'Did you have fun?'

'Yes, we did.'

'Fabulous. Tea is ready.'

'I thought maybe today *I* could feed him. Do his bath. Put him to bed.'

Any second now she would weep—Frederick looked as if a burden had been lifted from his shoulders.

'Great idea. In which case, if you don't mind, I've got a couple of errands to do. I'll be back for his bedtime.'

Frederick didn't need an observer—especially when Amil lobbed spaghetti Bolognese at him, as he no doubt would. And perhaps she could do something a little courageous as well. She needed to speak with Therese, the snooty seamstress, about her wedding dress.

Lycander tradition apparently had it that the royal seamstress had total input on the design of the dress, but surely the bride had a say as well. So maybe she would show Therese one of her designs.

Grabbing the relevant sketchbook, she blew Amil a kiss and allowed a cautious optimism to emerge as she made her way through the cavernous corridors to the Royal Sewing Room.

A knock on the door resulted in the emergence of one of Therese's assistants. 'Hi, Hannah, I wonder if Therese is around to discuss my dress.'

'She's popped out, but I know where the folder is. I'm sure she wouldn't mind you having a look.' Hannah walked over to a filing cabinet. 'I'll leave you to it.'

With that she scurried to an adjoining door and disappeared.

Sunita opened the folder and stared down at the picture—it was…was… Well, on the positive side it was classic—the designer a household name. On the negative side it was dull and unflattering. There didn't seem to be anything else in the folder, which seemed odd.

She headed to the door through which Hannah had disappeared, to see if there was another file, then paused as she heard the sound of conversation. She recognised Hannah's voice, and that of another assistant—Angela—and then the mention of her own name.

Of course she should have backed off there and then, taken heed of the old adage about eavesdroppers never hearing any good of themselves. But she didn't. Instead, breath held, she tiptoed forward.

'Do you think she knows?' asked Hannah.

'Knows what?' said Angela.

'That that's the dress Therese had designed for Lady Kaitlin—she was dead sure Lady K would marry Frederick. Everyone was, and everyone is gutted she didn't. Even that engagement ring—it was the one they had in mind for Kaitlin. Lady Kaitlin would have been a *proper* princess—everyone knows that. And she would have looked amazing in that dress—because it's regal and classic and not showy. As for Sunita—Frederick is only marrying her for the boy, which is dead good of him. He's a true prince. But Sunita can't ever be a true princess—she'll never fit in.'

Sunita closed her eyes as the flow of words washed over her in an onslaught of truth. Because that was what they were—words of truth. Otherwise known as facts. Facts that she seemed to have forgotten in the past weeks, somehow. She didn't know how she'd started to look at her marriage through rose-tinted glasses. How she had started to believe the fairy tale.

Fact: the sole reason for this marriage was Amil. Fact: Frederick's ideal bride would have been Lady Kaitlin or a woman of her ilk. Fact: fairy tales did not exist.

This must be what her mother had done—convinced herself that a handsome, charming English holidaymaker was her Prince, who would take her off into the sunset and a happy-ever-after. Perhaps that was what her father had done too—convinced himself that he could right past wrongs, that his family would welcome in his bastard child and everyone would live happily ever after.

Carefully she moved away from the door, leaving the folder on the desk, then picked up her sketchbook, and made her way out of the office, back along the marble floors, past the tapestry-laden walls, the heirlooms and antiquities collected over centuries, and back to her apartments.

She took a deep breath and composed her expression—this was a special day for Frederick and Amil and she would not spoil it.

Nor would she whinge and whine—there was no blame to be cast anywhere except at herself. Somehow she'd lost sight of the facts, but she wouldn't make that mistake again.

Entering the room, she halted on the threshold. Frederick sat on an armchair, Amil on his lap, looking down at a book of farm animals with intense concentration as Frederick read the simple sentences, and made all the noises with a gusto that caused Amil to chuckle with delight.

The book finished, Amil looked up and beamed at her and her heart constricted. Amil was the most important factor.

'Hey, guys. Looks like I'm back just in time.' Hard as she tried, she got it wrong—her voice was over-bright and a touch shaky, and Frederick's hazel eyes scanned her face in question.

'You OK?'

'I'm fine.' Walking over, she picked Amil up, hid her face under the pretext of a hug. 'How did tea and bath go?'

'Well, I have spaghetti down my shirt and bubbles in my hair, but we had fun, didn't we?'

'Abaadaaaaada!' Amil smiled and then yawned.

'I'll put him to bed.' Frederick rose and took Amil into his arms. Amil grizzled, but Frederick held on. 'Daddy's putting you to bed tonight, little fella. It'll be fine.'

And it was.

Fifteen minutes later Frederick emerged from Amil's bedroom, a smile on his face, headed to the drinks cabinet and pulled out a bottle of red wine.

Once they both had a glass in hand, he raised his. 'Thank you for today. You were right—I was afraid. Afraid I couldn't be a good parent... afraid I'd hurt him the way my parents hurt me. I thought doing nothing would be better than getting it wrong. Now I really hope that I can be a better parent than mine were, and can create a real bond with my son.'

The words made her happy—truly happy—and she wanted to step forward, to get close and tell him that, show him that happiness. But she didn't. Because close was dangerous—close had landed her in this scenario where she had distorted the facts with perilous consequence.

Perhaps it had been that magical physical intimacy in Goa, or maybe it had been a mistake to confide in him, to share her background and her fears and dreams. Whatever. No point in dwelling on the mistakes. Now it was vital not to repeat them.

So instead she stepped backwards and raised her glass. 'I'll drink to that. I'm so very pleased for you and Amil.'

And she was. But to her own horror, mixed into that pleasure was a thread of misery that she recognised as selfish. Because his love for Amil had never been in doubt—it had simply needed a shift in the dynamic of their relationship. A shift that had highlighted exactly what Sunita was—a by-product, a hanger-on, exactly as she been in her father's family. There only by default, by an accident of birth.

Well, she was damned if she would sit around here for the rest of her life being a by-product.

'Earth to Sunita?'

His voice pulled her out of her thoughts and she manufactured a smile, floundered for a topic of conversation. Her gaze fell on a folder—her plans for the state apartments.

'Would you mind having a look at this? I wanted your opinion before I went ahead.'

She picked up the file, opened it and pulled out the pictures, gazed down at them and winced. Every detail that she'd pored over so carefully screamed happy families—she'd done some of the sketches in 3D and, so help her, she'd actually imagined the three of them skipping around the place in some family perfect scene.

For their bedroom she'd chosen a colour scheme that mixed aquamarine blue with splashes of red. The double bed was a luxurious invitation that might as well have *bliss* written all over it.

She had to face it—her vision had included steamy nights with Frederick and lazy Sunday mornings with a brood of kids bouncing up and down. What had happened to her? This was a room designed with *love*.

This was a disaster. *Love*. She'd fallen in love. *Idiot, idiot, idiot.*

Frederick frowned. 'Are you sure you're OK?'

Reaching out, he plucked the pictures from her hand and she forced herself not to clutch onto them. *Think.* Before he looked at the pictures and figured it out. *Think.* Because even if he didn't work it out she couldn't share a room with him—

that would take intimacy to insane levels, and Frederick was no fool. He'd realise that she had done the unthinkable and fallen for him.

As he scanned the pictures desperation came to her aid. 'I wondered which bedroom you wanted.'

His head snapped up and his eyebrows rose. He placed the papers on the table. 'I assumed from these pictures that we would be sharing a room.'

Picking up her wine glass, she met his gaze. 'As I understood it, royal tradition dictates separate bedrooms and I assumed you'd prefer that.' Or at least she *should* have assumed that. 'However, obviously there will be some occasions when we *do* share—hence the design. If you want that bedroom I'll design the second one as mine. But if you don't then I'll take that one and we can discuss how you want yours to be.'

Stop, already. He'd got the gist of it and now she sounded defensive. Worse, despite herself, there was a hint of a question in the nuances of her tone, and a strand of hope twisted her heart. Hope that he'd take this opportunity to persuade her to share a room.

'What do you think?'

He sipped his wine, studied her expression, and she fought to keep her face neutral.

'I think this is some sort of trick question.'

'No trick. It's a simple need to know so I can complete the design. Also, as you know, I'll be giving an interview once the renovations are done, so it depends what you think the people of Lycander would prefer to see. We can pretend we share a room, if you think that would go down better, or...'

Shut up, Sunita.

Just because full-scale panic was escalating inside her, it didn't mean verbal overload had to implode. But she couldn't help herself.

'And then there is Amil to think about. I'm not sure that he should grow up thinking this sort of marriage is right.'

'Whoa! What is *that* supposed to mean? "This sort of marriage"? You say it like there's something wrong with it.'

'There *is* something wrong with it.'

It was almost as if her vocal cords had taken on a life of their own.

'Is this the sort of marriage you want Amil to have? An alliance? Presumably brokered by us? A suitable connection? Perhaps he will be lucky enough to get his own Lady Kaitlin. *Hell.*' She snapped her fingers. 'Perhaps we should get dibs on her first-born daughter.'

'Stop.' His frown deepened as he surveyed her expression. 'What is going on here? We agreed how our marriage would work—we agreed what we both wanted.'

'No, we didn't. I didn't want to get married at all. *You* did.'

'And you agreed.'

'Because there was no other choice.' She closed her eyes. 'There still isn't. But when Amil gets married I don't want it to be like this.'

For her son she wanted the fairy tale—she wanted Amil to love someone and be loved in return and live happily ever after. *The End.*

'There is nothing *wrong* with this.' His voice was urgent now, taut with frustration and more than a hint of anger. 'Amil will see two parents who respect each other, who are faithful to each other, who are polite to each other. There will be no uncertainty, no banged doors and no voices raised in constant anger. He'll have two parents who are there for him—I think he'll take that. God knows, *I* would have. And so would you.'

Touché. He was right. The problem was this wasn't about Amil. It was about her. *She* wanted the fairy tale. Her whole being cried out at the idea of a marriage of civility. Her very soul recoiled from the thought of spending the rest of her life tied to a man she loved who would never see her as more than the mother of his children—a woman he'd married through necessity not choice.

But she'd made a deal and she'd honour it. For Amil's sake. She wouldn't wrest Amil from his father, wouldn't take away his birthright. But neither would she stick around and moon in lovelorn stupidity. The only way forward was to kill love before it blossomed—uproot the plant now, before it sank its roots into her heart.

So she dug deep, conjured up a smile and said, 'You're right. I just had a mad moment.' She gave a glance at her watch. 'Anyway, don't you need to go? I thought you had a meeting about the casino bill?'

He hesitated. The frown still hadn't left his face. 'We'll talk in the morning.'

'Sure.'

Pride kept her cool, enigmatic smile in place as he turned and left the room. Then, ignoring the ache that squeezed her heart in a vicelike grip, she picked up her mobile.

'Harvey. It's me. We need to talk.'

The next morning Frederick approached the door to Sunita and Amil's apartments, forcing his steps to remain measured, forcing himself not to dwell on the previous night's conversation with Sunita. But her words still pummelled his conscience.

'I didn't want to get married. You did. There was no choice. There still isn't...'

The unpleasant edge of discomfort bit into him. *Shades of his father.*

He knocked on the door and entered, glanced around for Amil. Anxiety unfurled as he looked around and saw only Sunita at the table. 'Where's Amil?'

'With Gloria. I need to talk with you.'

Dressed in a simple white three-quarter-sleeved dress, belted with a striped blue and red band, she looked both elegant and remote. The only indication of nerves was the twist of her hands.

Sudden familiarity hit him—she'd had the same stance two years before, when she'd been about to leave. Panic grew inside him and he forced himself to keep still.

'Is anything wrong?'

'No. I've signed a modelling contract. Effective pretty much immediately. We need to discuss the details.'

'Immediately? But you said you wouldn't sign a contract straight away.'

'And I didn't. The first time the offer came in I refused. But now they have upped the ante and agreed to schedule the shoots around my commitments here. I'd be a fool to pass it up.'

A bleakness started to descend...a strange hollow pang of emptiness. 'What about Amil?'

'The first shoot is in India—in Mumbai. I'll

take Amil with me. We can stay with Nanni. Thereafter, whenever I can take him I will, or I will leave him in Lycander with you. Gloria will be here, and I will also ask Nanni if she can come and stay.'

She handed him a piece of paper with her schedule printed out.

Frederick frowned. 'This can't possibly have been arranged since last night.'

'No. Harvey was approached a few weeks ago—right after our first press release. The brand had dropped the model they had planned to use due to her lifestyle. They thought of me. I refused—but now I've changed my mind and luckily they still want me.'

'You didn't think to discuss this with me at all?'

'No. Any more than you discuss state business with *me*. Part of our marriage agreement was that I would resume my career as long as I fitted it around Lycander's needs. I understand some people may not approve, but to be honest with you a lot of people won't approve of me no matter what I do. The people wanted Kaitlin. I can't be her, or be like her. I will not try to fit into a box I'll never tick. I spent too many years doing that.'

Frederick knew there were things he should say, things he needed to say, but the words quite simply wouldn't formulate. All he wanted to do was tug her into his arms and beg her to stay—but that was not an option.

Because Sunita didn't want this marriage—she never had. He'd forced it upon her, caught her between a rock and a hard place. Self-disgust soured his tongue, froze his limbs. He was no better than his father. He had ridden roughshod over her wish not to marry him; he'd inveigled and manoeuvred and blazed down the trail he wanted regardless of her wishes. Worse, he'd bolstered himself with pious justifications.

But it was too late to stop the wedding, to release her from their marriage agreement—it would be an impossibility, the impact on Lycander, on Amil, too harsh.

Just Lycander? Just Amil? queried a small voice.

Of course. It made no difference to him. It *couldn't* make a difference to him

Why not?

The small voice was getting on his nerves now. It couldn't because he wouldn't let it. Over the past few weeks Sunita had very nearly slipped under his skin, and that way led to disaster, to pain and loss, to messy emotions that got in the way of a calm, ordered life. That would ruin this marriage before it even got underway.

The silence had stretched so taut now he could bounce off it. One of the terms of their marriage agreement *had* included her right to a career and he would not stand in the way. So he said, 'I understand. You're right to go.'

He looked down at the schedule again, but couldn't meet her gaze, couldn't seem to quell the spread of cold emptiness through his body and soul.

'I'll talk to Marcus—if you need to miss any functions I'll let him know it's OK. I know how important this contract is for you—anything I can do to help, I will.'

'Thank you. I appreciate that.'

He nodded, and yet still that desolation pervaded him, even as that unsquashable small voice exhorted him to *do* something. *Anything*.

But he couldn't. It wasn't in him. So instead he headed for the door.

One week later—Lycander Council Room

Frederick threw the pen across his desk and watched it skitter across the polished wood. Concentration wouldn't come. The words on the document blurred and jumped and somehow unerringly formed into images of Sunita. *Ridiculous*.

Shoving the wedge of paper away, he sighed,

and then looked up at the perfunctory knock on the door.

Seconds later he eyed his chief advisor in surprise. 'What's wrong?'

'Nothing. I thought you could do with a break. You've been closeted in here for hours. Days.'

'A break?' Frederick looked at Marcus blankly. 'Since when do you care about me having a rest?'

'Since you gave up on both sleep *and* food. So how about we grab a beer, shoot the breeze…?'

Frederick wondered if Marcus had perhaps already grabbed a few beers—though there was nothing in the other man's demeanour to suggest any such thing.

'Are you suggesting you and I go and have a beer?'

'Yes.'

'Why?'

Marcus shrugged. 'Why not?'

'Because I'm pretty damn sure you'd rather shoot yourself in the foot than have a beer with me.'

Dark eyebrows rose. 'Feeling tetchy?'

Damn right he was. Sunita had been gone for a week and his whole world felt…*wrong*…out of kilter. And he hated it. He loathed it that he didn't seem able to switch these emotions off. However hard he twisted the tap, they trickled on and on. Relentlessly.

'No. Just being honest. So, what gives?'

'I'm your chief advisor, right?'

'Right.'

'So here's some advice. Go after Sunita.'

'Sunita is in Mumbai on a photo shoot—why would I do that?'

'Because I think you love her.'

Frederick blinked, wondered if this conversation was a hallucination. 'Then you think wrong. You *know* all this—we're getting married because of Amil and for Lycander.'

'You can't kid a kidder. But, more importantly, why kid yourself?'

'With all due respect, has it occurred to you that this is a bit out of your remit?'

'Yes, it has.' Marcus started to pace the office, as he had so many times in the past year. 'But I'm talking to you now as Axel's best friend. I know Axel wouldn't want you to throw this away.'

Guilt and self-loathing slammed Frederick so hard he could barely stay upright. 'Hold it right there. You *don't* know what Axel would want.'

'Yes, I do. He'd want you to get on with life. *Your* life. Right now you're getting on with Axel's life, fulfilling his vision. I think Axel would want you to fulfil your own.'

Frederick searched for words before guilt choked him. He needed Marcus to stop—he couldn't listen to this any more.

'The accident… Axel's death was my fault. *I* should have been in that car. *I* was meant to go that state dinner. I passed it off to Axel.'

'I know.' Marcus's voice was measured. 'Axel was supposed to meet up with me that night. He told me you'd asked him to go because you had a party to go to. To celebrate a buy-out that no one thought you'd pull off.'

A buy-out he now wished he'd never tried for—if he could have pulled out any domino from the cause-and-effect chain he would. Most of all, though, he wished he hadn't chosen a party over duty.

He watched as Marcus continued to stride the floor. 'So why did you take this job with me? Why didn't you tell me you knew the truth?'

'I did what I believed Axel would have wanted. Axel was my best friend—we climbed trees as boys and we double-dated as young men. A few days before the accident he was trying to get up the courage to ask my sister on a date— he'd joke that he wished he had your charm. He cared about you very much, and he wouldn't have wanted you to punish yourself for the rest of your life.'

Marcus halted in front of the desk and leant forward, his hands gripping the edge.

'You didn't know what would happen—you didn't send Axel to his death.'

'But if I had chosen not to party, not to do what I wanted to do, then Axel would be alive now.'

Marcus shrugged. 'Maybe. But it didn't pan out like that. You can't turn the clock back, but you can make the most of your time now. I think you love Sunita, and if you do then you need to go for it—before you lose her. Axel's death should show you how life can change in a heartbeat—don't waste the life you've got. Axel wouldn't want it. And, for what it's worth, neither do I.'

Frederick stared at him. Emotions tumbled around him—poignant regret that Axel had never had the chance to ask Marcus's sister on a date, grief over the loss of his brother, a loss he had never allowed himself time to mourn, and gratitude that Marcus had given him a form of redemption.

'Thank you.'

There wasn't anything else he could say right now. Later there would be time. Time to grab those beers and sit and talk about Axel, remember him and mourn him. But now…

'Can I leave you at the helm? I need to go to Mumbai.'

'Good luck.'

Frederick had the feeling he'd need it.

Mumbai

Sunita smiled at Nanni across the kitchen, listened to the comforting whirr of the overhead fan, the sizzle of *malpua* batter as it hit the heated pan. Ever since Nanni had first made these sweet dessert pancakes Sunita had loved them.

'You don't have to make me breakfast every morning, Nanni, but I do so appreciate it."

'If I didn't you would eat nothing. And you do not need to stay in with me every night. I am sure there are parties and social events.'

'I'd rather be here.'

Totally true. She didn't want to socialise; she was too tired. Sleep deprivation, combined with the effort it took to work when all she wanted was to be back in Lycander. Irony of ironies, she wanted to be in a place that had rejected her, with a man who had rejected her. What a fool she was. But she'd be damned if anyone would know it—she'd dug deep, pulled up every professional reserve and hopefully pulled the wool over Nanni's eyes as well.

'You aren't happy.'

So much for the wool pulling endeavour.

Nanni put her plate in front of her; the scent of cardamom and pistachio drifted upward.

'Of course I am. The job is going great, I'm

back in Mumbai, Amil is here and I'm with my favourite grandmother.'

'And yet you still aren't happy. You can lie to me, Suni. But don't lie to yourself.'

'Sometimes you *have* to lie to yourself—if you do it for long enough the lie will become the truth.'

Or that was her theory. And, yes, there were holes and flaws in it, but it was a work in progress.

'That still doesn't make it *actual* truth,' Nanni pointed out with irrefutable logic. 'Do you no longer wish to marry Frederick?'

'I have no choice.'

'Yes, you do. There is always a choice—however hard a one it is. If you do not love him, don't marry him.'

'I *do* love him, Nanni. But he doesn't love me.' Tears threatened and she blinked them back.

'How do you know?'

'Because he has made it pretty clear.'

'By action or word?'

'What do you mean?'

'Love is not all about declarations—it is about demonstration. I told your mother how much I loved her repeatedly, but when the time came for me to show that, to fight for her, I failed miserably.'

'But you still loved her, Nanni, and she knew

that. She never blamed you—she blamed her father.'

'At least *he* had the courage of his convictions—he did what he did because he believed it to be right. I was a coward—I loved Leela, but not enough to fight for that love. It is one of the biggest regrets of my life, Suni—that I didn't fight for her, for *you*. So all I would say is think about your Frederick, and if there is a chance that he loves you then fight for that chance.'

Sunita stared down at her somehow empty plate and wondered if she were brave enough—brave enough to risk rejection and humiliation. And even if she was…

'I don't fit, Nanni. Even if he loved me I can't live up to Lady Kaitlin.'

'You don't need to live up to anyone—you just have to be yourself. Now, go,' Nanni said. 'Or you will be late for work.'

Frederick approached the site of the photo shoot, where the Gateway of India loomed in the twilight, its basalt stone lit up to create a magical backdrop, the turrets adding a fairy tale element for the photographer to take full advantage of.

The square was cordoned off for the shoot, and he joined the curious pedestrians who had stopped to watch. The people behind the cordon were packing up, so he slipped under the ropes, ignored the protest of a woman who approached

'Sir, I'm sorry, but…' There was a pause as she recognised him.

'I'm here to see Sunita.'

'Is she expecting you?'

'No. I thought I'd surprise her.'

'And you have.'

He spun round at the sound of Sunita's voice; his heart pounded, his gut somersaulted. She wore a square-necked red dress, in some sort of stretchy material that moulded her figure and fell to her knees in a simple drop.

'Is everything all right?'

'Yes. I want to…to talk.'

She hesitated and then nodded.

'OK. I'm done here. We can go by the wall over there and look out at the sea, if you like.'

'Perfect.'

'Are you sure everything is all right?' Concern was evident in her tone now, as she looked at him.

'Yes. No. I don't know.'

And he didn't—after all, this could be the most important conversation of his life and he could totally screw it up. He could lose her.

'I need to tell you something.'

'Actually, there is something I need to tell you as well.' Her hands twisted together and she looked away from him, out at the boats that bobbed on the murky water.

'Could I go first?' *Before he bottled it.*

She nodded.

'First I need to tell you about Axel. The accident that killed him—*I* should have been in the car. He took my place because I'd decided to go a party—a celebration of a buy-out deal. He not only took my place, he totally covered for me. He told everyone that it had been his idea, that he'd wanted to attend and I'd given up my place.'

Her beautiful brown eyes widened, and then without hesitation she moved closer to him. 'I am so very sorry. I cannot imagine what you went through—what you must still be going through. But please listen to me. It was *not* your fault—you did not know what would happen.'

'I know that, but...'

'But it doesn't help. I understand. I understand how many times you must think *if only* or *what if?* But you mustn't. I spent years thinking what if my mother hadn't died? What if she hadn't handed me over to my father? What if I could somehow have won his love? My stepmother's love. My sister's love? It made me question how I felt about my mother and it ate away at my soul—like this is eating away at yours. You can't know what would have happened. Axel might have died anyway. Your action wasn't deliberate—you wished Axel no harm.'

'That's what Marcus said.'

'He's a good man.'

'Yes.'

Frederick took a deep breath, looked at her beautiful face, her poise and grace, the compassion and gentleness and empathy in her gorgeous eyes.

'He said something else as well...'

Suddenly words weren't enough—he couldn't encompass how he felt in mere words. So instead he pushed away from the wall, and when she turned he sank to one knee.

'Frederick...?'

He could taste the sea spray, see the expression on her face of confusion, and hope soared in his heart as he took her left hand in his and removed the huge, heavy diamond—a ring chosen by someone else. He delved into his pocket and pulled out a box, purchased earlier from one of Mumbai's many jewellers.

'Will you marry me? For real. Not for Amil, not for Lycander, but because I love you. Heart, body and soul. Because I want to spend the rest of my life with you, wake up with you every morning. I want to live my life side by side with you. I want us to rule together, to laugh together, to live a life full of *all* the emotions. So, will you marry me? For real?'

His heart pounded and his fingers shook as he opened the box and took the ring out.

Her breathless laugh was caught with joy, the smile on her face so bright and beautiful his heart flipped.

'Yes. I *will* marry you. For real. Because I love you. That's what I wanted to tell you. That I love you. I didn't think in a million years that you could possibly love me back, but I wanted to tell you anyway. Heart, body and soul—they are all yours.'

He slipped the ring onto her finger and she lifted her hand in the air, watching the red and aquamarine stones interspersed with diamonds glint in the light of the setting sun.

'It's beautiful. Perfect.'

'I'm sorry about the other ring.'

'It makes this one all the more special. Did I mention I love it? Did I mention I love *you*?'

He rose to his feet. 'You did, but you can say it as many times as you like—those words won't ever get old.'

'No. Though Nanni says we have to back the words up. That love isn't only words—it's actions.' She stepped forward and looped her arms round his waist, snuggled in close. 'You *are* sure, aren't you? Sure this is real?'

'I have never been more sure of anything. I think I loved you from the start—I just couldn't admit it. Not to you, not to myself. You see, I didn't feel I deserved this joy.'

Her arms tightened around him. 'Because of Axel?'

'Partly. But even before that. I had parents who didn't give a damn—a mother who sold me for a crock of gold and a father who saw all his children as pawns or possessions. Love wasn't in the equation. Then I saw how much pain and angst love can cause—saw how losing Stefan tore Eloise apart. It was the same with Nicky, the twins' mother. My parents, who eschewed love, were happy and everyone else who *did* love was made miserable through that love. So I never wanted love to hold me hostage. I could see that emotions led to misery—that life was easier to control without emotion.

'Even two years ago you were different. But then you left, and Axel died, and I froze every emotion in order to cope. When you came back into my life all the emotions I'd bottled up for years kept surging up and I couldn't seem to shut them down any more. I didn't know what to do. I couldn't succumb to them because that was way too scary and it felt *wrong*. Axel died because of me. He'll never have the chance to live and love, have children, so how could it be right for me?'

'I am so very sorry about Axel, but I am sure he wouldn't have wanted you to give up on happiness.'

'I think you're right. Marcus seems sure of that too. And I know that if anything happened to me I wouldn't want *you* to shut your life down. I'd want you to live it to the full.'

'Nothing will to happen to you.' Her voice was fierce. 'And if it did I would never regret loving you.'

Frederick's heart swelled with the sheer wonder that he would share his life with this wonderful woman—and share it for real. The ups and the downs...everything.

He grinned down at her. 'I am so happy it doesn't seem possible. I never believed in my wildest dreams that you would love me too. I was willing to beg, fight—do whatever it took to persuade you to give me a chance to win your love.'

'You won that long ago but, like you, it took me a long time to admit it. I think deep down I knew the day I left you on the island with Amil. I couldn't have done that if I didn't trust you completely. And trust... I always thought that was for mugs and fools. My mother trusted my father once and ended up pregnant and abandoned. She trusted him again and it didn't end well for me. Fool me once, shame on you—fool me twice, shame on me. I figured it was best never to be fooled at all. Which meant the only way forward was never to trust anyone. But I trusted *you*. Even two years ago on some level I trusted you, or I would never have slept with you. But even after the island it all seemed so hard—you were so distant. And I felt like I used to—that I didn't fit. Everywhere I went people compared me to Kaitlin and...'

'You should have told me.'

'I couldn't. After all, you said that Kaitlin was your ideal bride.'

'I'm an idiot.' He cupped her face in his hands. 'I swear to you that you are the only bride I could ever want. Not because of Amil. Not because of Lycander. But because you are you and I *love* you. You make me whole. And if anyone makes any comparisons you send them to me.'

'No need. I've worked out where I've gone wrong—in my assumption that Kaitlin is better than me. She isn't—she is just different. I need to be a princess *my* way. Need to be myself.'

'And I know that means being a model—I will support your career every way I can.'

'About that... I'd better 'fess up. I'm not enjoying it one bit. I miss you and Amil...I miss Lycander. So I will fulfil this contract, but after that I will put a fashion portfolio together and send it off. And I also want to work for Lycander. I want to make it a fashion mecca—maybe set up a fashion show. One day we could rival Paris and London… There are so many options. But, whatever I decide, as long as you are by my side I know it will be OK.'

'Ditto.'

As he pulled her into his arms he knew this was the best alliance he could have ever made—because the only thing on the table was love.

* * * * *

If you enjoyed this story, check out these other great reads from Nina Milne

CLAIMED BY THE WEALTHY MAGNATE
THE EARL'S SNOW-KISSED PROPOSAL
RAFAEL'S CONTRACT BRIDE
CHRISTMAS KISSES WITH HER BOSS

All available now!

Get 2 Free Books,
Plus 2 Free Gifts—
just for trying the Reader Service!